# COLLABORATIVE LEARNING ACTIVITIES MANUAL

WITH ORIGINAL MATERIAL BY

## IRENE DOO

*Austin Community College*

# BASIC MATHEMATICS

## NINTH EDITION

# FUNDAMENTAL MATHEMATICS

## THIRD EDITION

# Marvin L. Bittinger

*Indiana University Purdue University at Indianapolis*

Addison Wesley

Boston San Francisco New York
London Toronto Sydney Tokyo Singapore Madrid
Mexico City Munich Paris Cape Town Hong Kong Montreal

# CONTENTS

## Activity 1.8    Make a budget for a road trip to your favorite destination.

| | |
|---|---|
| Focus | Problem solving and estimation |
| Time | 20–30 minutes |
| Group size | 3–4 |
| Materials | State and local highway maps for each group, calculators (optional) |
| Background | Planning a road trip involves several mathematical computations. For instance, the total distance to be traveled and the estimated cost for gas can be calculated using the concepts learned in this chapter. |

1. Before you begin your calculations, select a destination for a road trip you could take on long weekend. Decide on one destination for your group.

Origin: _____

Destination: _____

2. Using the appropriate state and local highway map(s), highlight the route you would take to get to your destination and back home again. Calculate the total distance you would need to drive. Round this distance to the nearest hundred.

Total distance: _____

Estimated distance: _____

3. Estimate the gallons of gas you would need for your trip. Use the miles per gallon (mpg) rating on one of your group member's vehicle. Then calculate the total cost of the gas. Use the price per gallon of gas in your area.

Gallons of gas needed: _____

Total cost for gas: _____

4.  Now decide how many days and nights it would take to complete the trip.  Then calculate the cost for the accommodations.

Days of travel: _____

Number of nights accommodation _____

Total cost for accommodations _____

5.  Based on the days of travel, calculate how many meals you would need to eat during the trip.  Then calculate the cost of the meals for all the people on this trip.

Number of meals per person: _____

Total number of meals: _____

Cost for meals: _____

6.  Summarize your estimated costs below.  Include a reasonable figure for the cost of miscellaneous items.  These might include the cost of admission tickets, souvenirs, parking, and tolls.

| Item | Estimated Cost |
|---|---|
| Gas: | |
| Accommodation: | |
| Meals: | |
| Miscellaneous: | |
| Total: | |

| | |
|---|---|
| Conclusion | As you can see, planning a budget for a road trip involves the operations of addition, subtraction, multiplication, and division, as well as estimation.  Use the steps given in this activity to plan a road trip for yourself and family or friends. |

## Activity 1.9    Use the order of operations as a group to simplify expressions.

| Focus | Order of operations |
|---|---|
| Time | 20–30 minutes |
| Group size | 3 |
| Background | Simplifying expressions using the rules for order of operations can be quite confusing for complicated expressions. Learning to simplify expressions as a group will help clarify the process. |

Rules for Order of Operations

| | | Do all calculations within parentheses before operations outside. |
|---|---|---|
| | **E** | Evaluate all exponential expressions. |
| | **MD** | Do all multiplications and divisions in order from left to right. |
| | **AS** | Do all additions and subtractions in order from left to right. |

1. Before you begin simplifying expressions, study the rules for order of operations above. Assign each group member to one of the steps listed. Write the name of the group member next to his or her assigned task in the table above. Note that the first step (calculations within parentheses) is not assigned. All group members will do this step together.

2. Now you are ready to simplify expressions as a group. Analyze the expression together and decide on the first step. If there are parentheses, decide whether the expression inside the parentheses needs to be simplified. Following the order of operations, **E** will perform his or her task before **MD**, and **MD** will perform his or her task before **AS**.

   Practice on the example on the next page. (This is example 10, Section 1.9 in your textbook.) The first step has been done for you: Add inside the parentheses. **AS** will do this step, writing "**AS**" in the left box, and writing the new expression below the original expression.

   Continue simplifying the expression by passing the problem to the appropriate group member for the next step. When you are done, compare your steps to those in Example 10, Section 1.9 in your textbook. If there are any discrepancies, discuss them within your group. Compare your result with the other groups. Are they the same? Discuss any differences with the other groups.

Example 10, Section 1.9

| | |
|---|---|
| | $4^2 \div (10 - 9 + 1)^3 \cdot 3 - 5$ |
| **AS** | $4^2 \div (1 + 1)^3 \cdot 3 - 5$ |
| | |
| | |
| | |
| | |
| | |
| | |

3.  Once you understand the process, choose an expression from Exercise Set 1.9 in your textbook to simplify as a group. Use the table on the next page to organize your work. Make as many copies as you need. Alternatively, you can draw the table on a blank sheet of paper.

    Do as many problems as you can in the time allotted. Make sure you choose at least one of the more complicated expressions from Exercises 57-66.

| Conclusion | This activity should help you gain a better understanding of the rules for order of operations. You can also use this group method when you encounter the order of operations in Sections 3.7, 4.4, and 10.5. For your convenience, there are separate activities for these sections. See the table of contents. |
|---|---|

Original expression _____

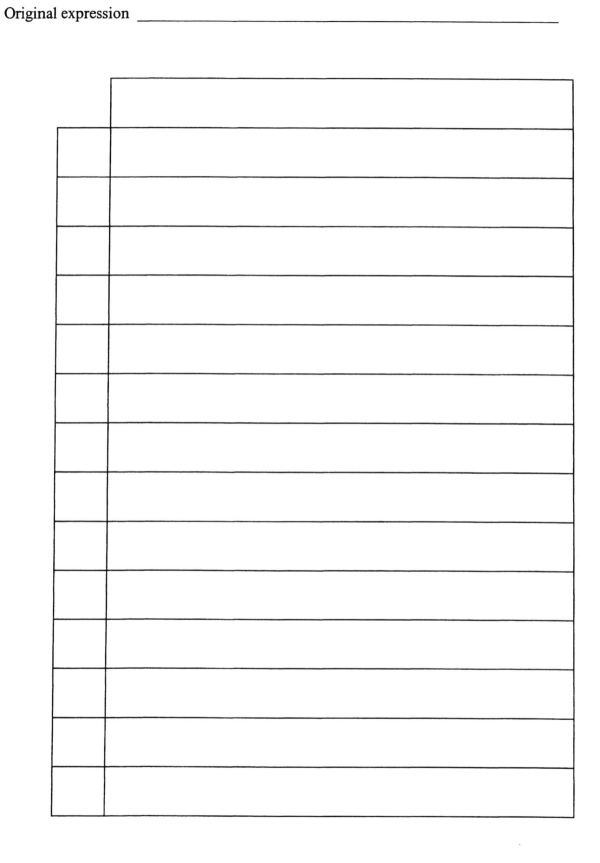

**Activity 2.1    Find all the prime numbers less than 100, using the Sieve of Eratosthenes.**

| | |
|---|---|
| Focus | Prime and composite numbers |
| Time | 10–15 minutes |
| Group size | 2 |
| Materials | Colored pencils (optional) |
| Background | One of the methods for finding prime numbers was developed around 200 BC by a mathematician named Eratosthenes. He used the process of elimination to "sift" out the composite numbers, leaving only prime numbers. His method became known as the Sieve of Eratosthenes. |

1.    In Section 2.1 of your textbook, a prime number is defined as a natural number that has exactly two different factors, itself and 1. For example, the number 7 is prime because it has only the factors 1 and 7. The number 14, on the other hand, is not prime because 7 is a factor of 14. Looking at the definition from another point of view, any number that is a multiple of another number will not be prime. In the example above, 14 is a multiple of 7, and so 14 is not prime.

In this activity, you will cross off all multiples of prime numbers from a grid of numbers. When you are done, the remaining numbers will be prime.

2.    Look at the grid on the next page. The number 1 has already been crossed off, as it is not a prime number. The smallest number that is not crossed off is 2. Begin by circling the number 2 on the grid. Then, list the first 10 multiples of 2 in the space below:

Now, cross off these numbers from the grid. You may want to use a colored pencil to cross off the numbers. Continue crossing off multiples of 2 until you reach the end of the grid.

3.    Next, look for the smallest number that is not crossed off and circle it. This is the next prime number. List the first 10 multiples of this number in the space below:

Cross off these numbers from the grid. Continue, as before, crossing off multiples of the number until you reach the end of the grid.

| X | 2 | 3 | 4 | 5 | 6 | 7 | 8 | 9 | 10 |
|---|---|---|---|---|---|---|---|---|----|
| 11 | 12 | 13 | 14 | 15 | 16 | 17 | 18 | 19 | 20 |
| 21 | 22 | 23 | 24 | 25 | 26 | 27 | 28 | 29 | 30 |
| 31 | 32 | 33 | 34 | 35 | 36 | 37 | 38 | 39 | 40 |
| 41 | 42 | 43 | 44 | 45 | 46 | 47 | 48 | 49 | 50 |
| 51 | 52 | 53 | 54 | 55 | 56 | 57 | 58 | 59 | 60 |
| 61 | 62 | 63 | 64 | 65 | 66 | 67 | 68 | 69 | 70 |
| 71 | 72 | 73 | 74 | 75 | 76 | 77 | 78 | 79 | 80 |
| 81 | 82 | 83 | 84 | 85 | 86 | 87 | 88 | 89 | 90 |
| 91 | 92 | 93 | 94 | 95 | 96 | 97 | 98 | 99 | 100 |

4.  Repeat step 3 until all multiples are crossed off. The circled numbers are the prime numbers less than 100. Write the list of circled numbers in the space below:

5.  Compare this list with the table of primes given in section 2.1 of your textbook. Are there any differences between the lists? If there are, check your grid to see if you crossed off all multiples. Check also that you did not accidentally cross off a number that is not a multiple.

| Conclusion | The Sieve of Eratosthenes can be used anytime you need to list the first few prime numbers. For example, if you need all the prime numbers up to 50, make a list of the numbers from 1 to 50, and start crossing out the multiples of 2, 3, 5, etc. |
|---|---|

## Activity 2.2    Use the divisibility rules and properties of numbers to discover an unknown number.

| | |
|---|---|
| Focus | Rules for divisibility, place value |
| Time | 20–30 minutes |
| Group size | 2 |
| Background | The rules for divisibility given in Section 2.2 of your textbook provide you with fast ways of determining whether numbers are divisible by 2, 3, 4, 5, 6, 8, 9, and 10. This activity will provide practice with these rules, as well as experience in problem solving. |
| Instructor notes | Make transparencies of Puzzles A and B. In step 4, show the puzzles using an overhead projector, and use a sheet of paper to reveal clues one at a time. You can find more puzzles in the book *Logic Number Problems*, available from Dale Seymour Publications. |

For your convenience, the divisibility rules from Section 2.2 are repeated here.

| | |
|---|---|
| 2 | A number is divisible by 2 (is even) if it has a ones digit of 0, 2, 4, 6, or 8 |
| 3 | A number is divisible by 3 if the sum of its digits is divisible by 3 |
| 4 | A number is divisible by 4 if the number named by its last two digits is divisible by 4 |
| 5 | A number is divisible by 5 if its ones digit is 0 or 5 |
| 6 | A number is divisible by 6 if its ones digit is 0, 2, 4, 6, or 8 (is even) and the sum of its digits is divisible by 3 |
| 8 | A number is divisible by 8 if the number named by its last three digits is divisible by 8 |
| 9 | A number is divisible by 9 if the sum of its digits is divisible by 9 |
| 10 | A number is divisible by 10 if its ones digit is 0 |

1. Each puzzle in this activity gives you clues to the value of an unknown number. The objective is to determine the unknown number by using the fewest number of clues. The clues will be given to you one at a time.

2. First, practice on the following set of clues. Read the clues one at a time, using a sheet of paper to cover up the clues further down.

| | Clue | Possible solution(s) | Reasoning |
|---|---|---|---|
| 1 | It is a 3-digit number | __ __ __ | Write one blank for each digit |
| 2 | It is divisible by 5 | __ __ 0<br><br>__ __ 5 | To be divisible by 5, the last digit must be 0 or 5 |
| 3 | It is an even number | __ __ 0 | The last digit must be 0, 2, 4, 6, or 8 |
| 4 | It is less than 400 | 3 __ 0<br><br>2 __ 0<br><br>1 __ 0 | The hundreds digit must be less than 4 |
| 5 | Each digit is different | 3 2 0 | No digit can be repeated |
| 6 | Its tens digit is greater than its ones digit | 3 1 0<br><br>2 1 0 | The tens digit must be 1 or higher, and the hundreds digit must be 2 or higher. |
| 7 | Its hundreds digit is greater than its tens digit | | |
| 8 | It is divisible by 3 | 2 1 0 | The sum of its digits must be divisible by 3. |
| 9 | It has only one odd digit | | These clues confirm that the number is 210. They are actually not needed to solve the puzzle. |
| 10 | Its tens digit is 1 | | |

Notice that some clues must be considered together (clues 5, 6, and 7), and that only the first 8 clues are needed to solve this puzzle.

3. Here's another puzzle to practice on. One group member writes down the possible solutions, as was done in the example on the previous page. Use complete sentences when writing the reasons for each possible solution. Read the clues one at a time, using a sheet of paper to cover up the clues further down.

| | Clue | Possible solution(s) | Reasoning |
|---|---|---|---|
| 1 | It is a 3-digit number | | |
| 2 | It is an odd number | | |
| 3 | One of the digits is 7 | | |
| 4 | It is divisible by 5 | | |
| 5 | It is less than 700 | | |
| 6 | It has no even digits | | |
| 7 | It is divisible by 3 | | |
| 8 | It is greater than 200 | | |
| 9 | Each digit is different | | |
| 10 | It is a multiple of 25 | | |

When you are done, compare your group's result with the results of the other groups in your class. How many clues did your group need to solve this puzzle? Could you have determined the unknown number with fewer clues? Did you use the remaining clues (if any) to check your answer?

4.  Now, let's add a little competition to the problem-solving process. Each group will work as a team to solve a puzzle. Take turns, so each group member has a chance to do the writing. Your instructor will reveal the clues one at a time. The goal is to be the first group to correctly deduce the unknown number by using the fewest number of clues. Your instructor will discuss the scoring scheme; alternatively, the class can propose a scheme that is acceptable to all. The scoring scheme should take into account the correctness of the number, the penalty for a wrong number, the number of clues used, and the penalty for using more clues than needed.

| Conclusion | This activity should help you gain experience in applying the divisibility rules. As a side benefit, the problem solving techniques used in solving the puzzles will be useful in solving the applications in your textbook. |
| --- | --- |

# Puzzle A

| | |
|---|---|
| 1 | It is a 3-digit number |
| 2 | It is divisible by 5 |
| 3 | It is an odd number |
| 4 | Each of its digits is different |
| 5 | Its tens digit is less than its ones digit |
| 6 | Its hundreds digit is less than its tens digit |
| 7 | It is greater than 200 |
| 8 | It is divisible by 3 |
| 9 | It has two odd digits |
| 10 | Its tens digit is 4 |

# Puzzle B

| | |
|---|---|
| 1 | It is a 3-digit number |
| 2 | It is divisible by 5 |
| 3 | Its hundreds digit is 8 |
| 4 | It is divisible by 3 |
| 5 | Its tens digit is less than its ones digit |
| 6 | None of its digits are repeated |
| 7 | The sum of two of its digits is 10 |
| 8 | It has only one odd digit |
| 9 | It is divisible by 11 |
| 10 | Its tens digit is 2 |

## Activity 2.5    Use fraction bars to represent equivalent fractions.

| | |
|---|---|
| Focus | Equivalent fractions |
| Time | 10–15 minutes |
| Group size | 3 |
| Background | Equivalent fractions are used extensively when adding, subtracting, and simplifying fractions. In Section 2.5 of your textbook, the process of multiplying by one is used to find equivalent fractions, and to simplify fractions. This activity will give you a better understanding of these concepts. |
| Instructor notes | Each group will need one set of bars from the next three pages. Copy the pages on card stock, and cut along the heavy outline of each bar; do not cut within each fraction bar. You can also purchase fraction bars. |

1.  Two fractions are equivalent if they represent the same number. For example, 2/3 and 4/6 are two names for the same number. We will use fraction bars to show how equivalent fractions can be represented visually. Take your group's set of fraction bars and mix them up. One group member selects the bar that represents 2/3, and places it in the box below.

$$\frac{2}{3}$$

$$\frac{4}{6}$$

Taking turns, the next group member selects the fraction bar that represents 4/6, and places it in the appropriate box above. The entire group should then examine the two bars. Are the fractions equivalent? Compare the shaded areas in each bar. Are the areas the same? Pick another bar that has the same shaded area as 2/3 and 4/6 and place it in the last box on the previous page. Write the name for the fraction to the right of the box. Are the fractions equivalent?

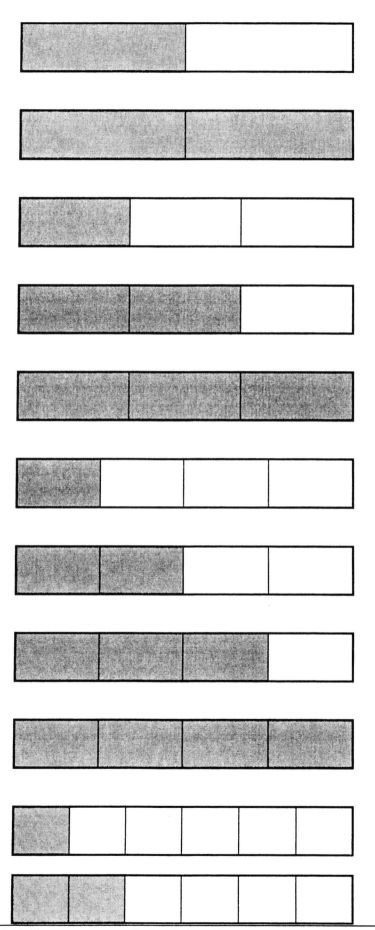

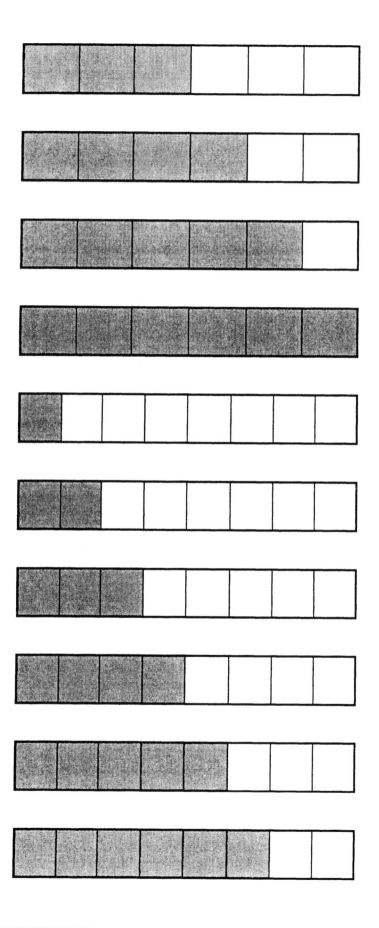

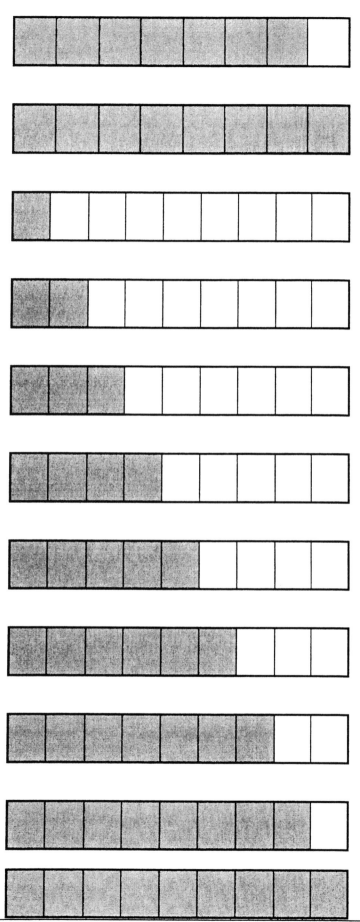

2.   Next, one group member picks the bar that represents 1/2 and place it in the appropriate box below.

$$\frac{1}{2}$$

Taking turns, choose three other bars that are equivalent to 1/2 and place them in the boxes above.  Write their fraction names to the right of the appropriate box.

Study the fraction names for these equivalent fractions.  How are the numerators and denominators related?  Use complete sentences to write your answer.

Write fractions equivalent to 1/2, but with denominators specified.  Check your answers with the other group members, and discuss any answers that do not match.  If necessary, consult with another group until an agreement is reached on the correct answers.

$$\overline{10} \qquad \overline{18} \qquad \overline{24} \qquad \overline{100}$$

3.  Next, consider the fraction 3/9.  In turn, each group member should look through the pile of fraction bars, and find the bar that represents 3/9, and two other bars that have the same shaded area as 3/9.  Place them in the boxes below, and write the fraction names.  Which fraction is simplest?  Why?

$$\frac{3}{9}$$

4.  Finally, one group member should draw the bar that represents 9/12.

Another group member then simplifies 9/12 using the process of multiplying by one, as shown in Section 2.5 of your textbook.  Write the simplified fraction below, and the third group member should find the fraction bar that represents it.

Compare this fraction bar with the one drawn for 9/12.  Are the shaded areas the same?  Why or why not?  Be sure to use complete sentences for your answer.

| Conclusion | Equivalent fractions are represented by the same shaded area on a fraction bar. Therefore, to find equivalent fractions, you multiply by a form of 1, as described in Section 2.5 of your textbook.  Conversely, to write a fraction in simplest form, you would reverse the process, and remove a factor of 1. |

## Activity 3.1    Find the least common multiple of two or more numbers using shaped markers.

| | |
|---|---|
| Focus | Least Common Multiples |
| Time | 20–30 minutes |
| Group size | 2 |
| Background | The textbook describes two methods for finding the least common multiple (LCM) of a set of numbers: using multiples and using factorizations. To find the LCM using factorizations, first, find the prime factorization of each number; then, create a product of factors, using each factor the greatest number of times it occurs in any one factorization. The second part of the factorization method requires the creation of a product of factors. We will see how these factors are chosen by using shaped markers to represent the factorizations. This visualization should give you a clearer picture of the process. |
| Instructor notes | Copy the next page on card stock, and cut out the markers. Each group will need one set of markers. You can also purchase sets of markers (also called pattern blocks). |

1.    Study the set of markers; notice that each type of marker represents a different prime number, as follows.

| Marker | Prime number |
|---|---|
| Circle | 2 |
| Triangle | 3 |
| Square | 5 |
| Hexagon | 7 |

For the first part of this activity, use a restricted subset of markers consisting of 3 circles, 3 triangles, and 3 squares.

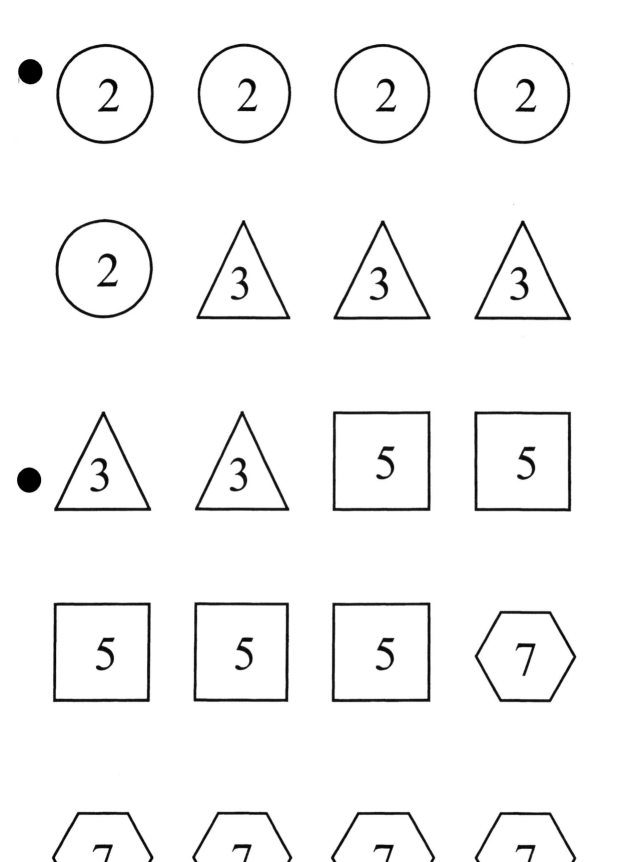

2. Use the restricted set of markers to represent each of the following prime factorizations. You will need to place all used markers back in the subset before you work on the next prime factorization. Take turns selecting the markers, so each group member gets to practice this step.

$$90 \; = \; \underline{\quad 2 \quad} \; \cdot \; \underline{\quad 3 \quad} \; \cdot \; \underline{\quad 3 \quad} \; \cdot \; \underline{\quad 5 \quad}$$

$$150 \; = \; \underline{\quad 2 \quad} \; \cdot \; \underline{\quad 3 \quad} \; \cdot \; \underline{\quad 5 \quad} \; \cdot \; \underline{\quad 5 \quad}$$

Notice that by replacing markers back into the subset, you were able to reuse some markers. Therefore, it may not have been necessary to have 3 of each kind of marker. Experiment with your partner to find the minimum number of each marker needed to represent the prime factorizations given above. Remember that you do not need to represent the factorizations at the same time. List the minimum set of markers below.

5. Now, use the full set of markers and practice finding the minimum set of markers needed for each of the following prime factorizations. Represent the minimum set with markers, and also write the prime factorization of the minimum set. Each group member should do one problem individually, then exchange papers and check each other's work.

| | Problem | Minimum set of markers | Prime factorization of minimum set |
|---|---|---|---|
| A. | $18 = 2 \cdot 3 \cdot 3$ <br> $30 = 2 \cdot 3 \cdot 5$ | | |

| | | |
|---|---|---|
| B.     $75 = 3 \cdot 5 \cdot 5$ <br>         $70 = 2 \cdot 5 \cdot 7$ | | |

6. The minimum set of markers needed for a group of prime factorizations is the least common multiple (LCM) of the group of numbers. Find the LCM of the following groups. You may use the markers to help you find the LCM.

| Problem | LCM |
|---|---|
| A.     50 <br>         75 | |
| B.     24 <br>         18 | |

| Conclusion | The factorization method for finding least common multiples can be visualized using prime number markers. The minimum set of markers needed to represent the prime factorization of each number will give the LCM of the numbers. |
|---|---|

**Activity 3.3    Arrange sockets and drill bits in fractional sizes from smallest to largest.**

| | |
|---|---|
| Focus | Order of fractions |
| Time | 15–20 minutes |
| Group size | 3 |
| Materials | Socket set (optional), drill bit set (optional) |
| Background | In the United States, tools are typically classified using either English or metric units. For example, socket sets are sold in fractional sizes (fractions of an inch), or metric sizes (millimeters). When fractional sizes are used, it is important to be able to compare the relative size of each tool by comparing the fractions. |
| Instructor notes | Copy the page of socket and drill bit sets on card stock, and cut out them out. Each group will need one set of each type. You can also ask the students to bring socket sets and drill bits to class. |

1.   Each group member takes one of the sets and arranges the tools in order from smallest to largest. Write down your result below.

     Set _____

     _____

     smallest                                                                    largest

2.   Now, pass your paper to the group member to your right and check the result from step 2 on the paper you receive. If you disagree with the result, discuss this with the paper's owner and try to resolve any discrepancies.

3.   Next, we will work with the drill bit sets (Sets B and C) only. Mix up the drill bits in the two sets; your group should have a total of 30 drill bits. Deal 10 bits for each group member. One group member starts by choosing one of his or her bits, and placing it on the table. The next group member to the right chooses one of his or her drill bits, and places it next to the bit already on the table such that the smaller bit is to the left. He or she will need to decide which bit is smaller. Continue in turn, with each group member choosing a drill bit from his or her stack, and placing it in the correct position by the other bits already on the table.

4.  When all group members have finished placing their drill bits on the table, write down the result below.

5.  Compare your group's result with the results of the other groups. Resolve any discrepancies by discussing them with each other.

6.  The final step in this activity is to rewrite each fraction in step 5 in terms of the least common denominator of all the fractions. Multiply each fraction by 1, using the appropriate notation, $n/n$. Write the result below.

7.  Compare the lists in steps 5 and 7. Which list is easier to read? Why? Use complete sentences in your answer.

The list in step 5 is used by manufacturers to classify drill bits. Why do you suppose a manufacturer would choose the list in step 5 over the list in step 7?

| Conclusion | Fractions in everyday life are typically given in simplified form. Thus, it is important to be able to compare fractions with different denominators. Remember to multiply by 1 to make the denominators the same. The skills learned from this activity can help you do this. |
| --- | --- |

## SET A:  SOCKETS

| 3/8 | 13/16 | 9/32 | 5/8 | 3/16 | 1/4 | 11/16 | 3/4 | 15/16 | 9/16 |
|-----|-------|------|-----|------|-----|-------|-----|-------|------|
| 1/2 | 11/32 | 7/32 | 5/16 | 7/8 | 7/16 | | | | |

## SET B:  SMALL DRILL BITS

| 15/64 | 5/32 | 1/4 | 5/64 | 3/32 | 3/16 | 7/32 | 13/64 | 1/8 | 7/64 |
|-------|------|-----|------|------|------|------|-------|-----|------|
| 9/64 | 1/16 | 11/64 | | | | | | | |

## SET C:  BIGGER DRILL BITS

| 13/32 | 23/64 | 7/16 | 1/4 | 9/32 | 19/64 | 5/16 | 17/64 | 3/8 | 29/64 |
|-------|-------|------|-----|------|-------|------|-------|-----|-------|
| 1/2 | 25/64 | 11/32 | 21/64 | 27/64 | 15/32 | 31/64 | | | |

## Activity 3.5     Add and subtract mixed numerals using fraction strips.

| | |
|---|---|
| Focus | Addition and subtraction of mixed numerals |
| Time | 10–15 minutes |
| Group size | 3 |
| Background | Addition and subtraction of mixed numerals involves the concepts of equivalent fractions, borrowing, and converting from fractional notation to mixed numerals. This activity is designed to help you visualize these concepts by using fraction strips to represent the mixed numerals. |
| Instructor notes | Copy the next page onto card stock, and cut out the fraction strips. Each group will need three sets of strips. Cut out each fraction strip, and also cut along the black lines inside each strip. You can also purchase fraction strips. |

1.    One group member will be the banker. This person keeps all the fraction pieces, and hands pieces to the other two group members at the appropriate times. The other two group members will take the roles of the top and bottom players in the addition and subtraction problems. The top player will be responsible for the first (or top) number in the problem, while the bottom player takes care of the second (or bottom) number. Write the names of each group member next to his or her designated role.

Banker     _____

Top Player     _____

Bottom Player     _____

| 1 |
|---|

| 1 |
|---|

| 1 |
|---|

| 1 |
|---|

| 1 |
|---|

| $\frac{1}{2}$ | $\frac{1}{2}$ |
|---|---|

| $\frac{1}{3}$ | $\frac{1}{3}$ | $\frac{1}{3}$ |
|---|---|---|

| $\frac{1}{4}$ | $\frac{1}{4}$ | $\frac{1}{4}$ | $\frac{1}{4}$ |
|---|---|---|---|

| $\frac{1}{6}$ | $\frac{1}{6}$ | $\frac{1}{6}$ | $\frac{1}{6}$ | $\frac{1}{6}$ | $\frac{1}{6}$ |
|---|---|---|---|---|---|

| $\frac{1}{8}$ | $\frac{1}{8}$ | $\frac{1}{8}$ | $\frac{1}{8}$ | $\frac{1}{8}$ | $\frac{1}{8}$ | $\frac{1}{8}$ | $\frac{1}{8}$ |
|---|---|---|---|---|---|---|---|

| $\frac{1}{12}$ | $\frac{1}{12}$ | $\frac{1}{12}$ | $\frac{1}{12}$ | $\frac{1}{12}$ | $\frac{1}{12}$ | $\frac{1}{12}$ | $\frac{1}{12}$ | $\frac{1}{12}$ | $\frac{1}{12}$ | $\frac{1}{12}$ | $\frac{1}{12}$ |
|---|---|---|---|---|---|---|---|---|---|---|---|

3. Give all the fraction pieces to the banker. We will begin by practicing on the following problem from Section 3.5 in your textbook.

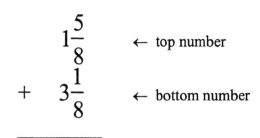

$$1\frac{5}{8} \qquad \leftarrow \text{top number}$$
$$+ \quad 3\frac{1}{8} \qquad \leftarrow \text{bottom number}$$

First, the banker hands one $\boxed{1}$ piece and five $\boxed{\frac{1}{8}}$ pieces to the top player; similarly, the bottom player gets three $\boxed{1}$ pieces and one $\boxed{\frac{1}{8}}$ piece. The two players then decide if the pieces can be added. Remember that only pieces of the same size can be combined. In this example, the $\boxed{1}$ pieces can be added together, and the $\boxed{\frac{1}{8}}$ pieces can also be added. Write the mixed numeral represented by the sum of all the pieces.

Now, consider the $\boxed{\frac{1}{8}}$ pieces. Can they be exchanged for larger pieces? If yes, make the change with the banker. What pieces do you now have? List them below.

| Piece type | Number of pieces |
| --- | --- |
|  |  |

Write the mixed numeral represented by the pieces. This will be the answer to the addition problem.

4. The next example will show you how to exchange pieces to get pieces of the same size.

$$1\dfrac{2}{3}$$
$$+\quad 3\dfrac{5}{6}$$
_____

As before, the banker hands the appropriate pieces to the two players. However, this time the $\boxed{\dfrac{1}{3}}$ pieces and the $\boxed{\dfrac{1}{6}}$ pieces cannot be added because they are of different sizes. Decide within your group how to exchange pieces with the banker so that both players end up with pieces of the same size.

Now, add the pieces and write your result below.

| Piece type | Number of pieces |
|---|---|
|  |  |

What mixed numeral is this?

See if you can exchange any pieces for a larger size piece. The goal is to have as few pieces as possible; exchanging smaller sized pieces for larger ones. What is your final mixed numeral? Compare your result with the other groups in the class. Do you all have the same answer? Discuss any differences with the other groups.

5. Next, use the fraction pieces to work the following problems.  Switch roles within your group so that each group member gets to play the different roles.

$$3\frac{3}{4}$$
$$+\quad 2\frac{1}{2}$$
_____

| RESULT | |
|---|---|
| Piece type | Number of pieces |
| | |

$$2\frac{5}{8}$$
$$+\quad 1\frac{1}{6}$$
_____

| RESULT | |
|---|---|
| Piece type | Number of pieces |
| | |

$$3\frac{1}{4}$$
$$+\quad 2\frac{5}{6}$$
_____

| RESULT | |
|---|---|
| Piece type | Number of pieces |
| | |

6. Let's now look at subtraction of mixed numerals. Consider the following problem.

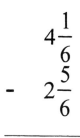

The top player gets four $\boxed{1}$ pieces, and one $\boxed{\frac{1}{6}}$ piece, while the bottom player gets two $\boxed{1}$ pieces and five $\boxed{\frac{1}{6}}$ pieces. Since this is a subtraction problem, the bottom player must take pieces from the top player to match the pieces he or she has.

In this example, the bottom player needs to get two $\boxed{1}$ pieces and five $\boxed{\frac{1}{6}}$ pieces from the top player. The top player gives two of his or her $\boxed{1}$ pieces to the bottom player. However, there are not enough $\boxed{\frac{1}{6}}$ pieces, so the top player must exchange one $\boxed{1}$ piece for six $\boxed{\frac{1}{6}}$ pieces from the banker. Then, he or she can give five $\boxed{\frac{1}{6}}$ pieces to the bottom player.

The pieces remaining with the top player will be the answer to this problem. Write the result below.

| Piece type | Number of pieces |
|---|---|
| | |

What mixed numeral is this? When possible, you should exchange smaller sized pieces for an equivalent number of larger size pieces.

7. Now that you have some experience working with fraction pieces, do the following subtraction problems. Exchange pieces with the banker to get the same size pieces. You may also need to exchange a whole piece for the equivalent number of fraction pieces so you can subtract. When you have a result, don't forget to check if smaller sized pieces can be exchanged for an equivalent number of larger sized pieces.

Switch roles after doing a problem, so each group member has a chance to play each role. Check your results with the other groups in the class.

$$2\frac{1}{8}$$

$$-\ 1\frac{1}{2}$$

| RESULT | |
|---|---|
| Piece type | Number of pieces |

$$3\frac{5}{8}$$

$$-\ 1\frac{3}{4}$$

| RESULT | |
|---|---|
| Piece type | Number of pieces |

$$4\frac{1}{6}$$

$$-\ 2\frac{5}{8}$$

| RESULT | |
|---|---|
| Piece type | Number of pieces |

| Conclusion | Working with the fraction pieces should give you a better insight into the operations of addition and subtraction of mixed numerals. This activity is especially useful for visualizing the borrowing step in subtraction. |
|---|---|

## Activity 3.6    Analyze stock market prices.

| | |
|---|---|
| Focus | Operations with mixed numerals |
| Time | 20–25 minutes |
| Group size | 3 |
| Background | The price of a share in the stock market is usually given as a mixed numeral, like $46\frac{3}{4}$. The denominators of these fractions are usually 2, 4, 8, or 16. When buying or selling stocks, you will need to be able to add, subtract, multiply, and divide mixed numerals. This activity will give you some practice with these calculations. |

1.   As a group, study the price history of the following stock. This stock is traded on the NASDAQ (National Association of Stock Dealers' Automated Quotation) stock exchange.

| 3COM | | |
|---|---|---|
| | Price* | Change* |
| Monday | $46\frac{3}{4}$ | $+3\frac{1}{2}$ |
| Tuesday | $51\frac{1}{8}$ | $+4\frac{3}{8}$ |
| Wednesday | $47\frac{13}{16}$ | $-3\frac{5}{16}$ |
| Thursday | $48\frac{1}{2}$ | $+\frac{11}{16}$ |
| Friday | $49\frac{1}{16}$ | $+\frac{9}{16}$ |

*The prices given are "closing prices," which is the price at the close of the business day. For example, Monday's price of $46\frac{3}{4}$ is the price at 5 pm Eastern time on Monday. The price during the day can fluctuate up or down. For this activity, we will assume that the price is valid for the entire day. The change column tells you if the price of the stock increased (+) or decreased (−).

The change column is calculated by subtracting the previous day's price from the current day's price. For example,

Tuesday's change = Tuesday's price – Monday's price

Verify the change value for Tuesday by doing the calculation.

Is it possible to check the change value for Monday? Why or why not?

Now, suppose you are an investor, and you wish to purchase 200 shares of 3COM on Monday. Calculate the cost of this purchase.

If you sell the 200 shares on Thursday, how much money will you get? Will you make a profit or suffer a loss? By how much?

2.  Each group member should pick one of the stocks listed on the next page and answer the following questions.

    a)  Calculate the missing prices and change values of the stock you chose. When you are done, exchange papers with the others in your group and check each other's work.

    b)  Calculate the cost of purchasing 200 shares of your stock on Monday.

    c)  If you sell the 200 shares on Friday, how much money will you receive? Will you make a profit or suffer a loss? By how much?

|  | IBM | | Dell | | Compaq | |
|---|---|---|---|---|---|---|
|  | Price | Change | Price | Change | Price | Change |
| Monday | $168\frac{3}{4}$ | $-1\frac{1}{4}$ | $97\frac{1}{2}$ | $+3\frac{1}{4}$ | $94\frac{5}{8}$ | $+\frac{5}{8}$ |
| Tuesday | $173\frac{3}{4}$ | | | $+2\frac{1}{8}$ | $97 | |
| Wednesday | $175\frac{1}{8}$ | | | $+7\frac{1}{8}$ | | $+5\frac{3}{4}$ |
| Thursday | $172\frac{1}{2}$ | | | $+\frac{13}{16}$ | | |
| Friday | $173\frac{3}{8}$ | | | $-\frac{5}{16}$ | $105\frac{1}{8}$ | $+4\frac{1}{8}$ |

IBM's and Compaq's stock are traded on the New York Stock Exchange (NYSE).

Dell's stock is traded on the NASDAQ stock exchange.

| Conclusion | This activity gives you hands-on experience with a real-world application of mixed numeral arithmetic. For further practice, you can use the newspaper to track the stock prices of several stocks that are currently traded. |
|---|---|

**Activity 3.7    Use the order of operations as a group to simplify expressions.**

| Focus | Order of operations |
|---|---|
| Time | 20–30 minutes |
| Group size | 3 |
| Background | Simplifying expressions using the rules for order of operations can be quite confusing for complicated expressions. Learning to simplify expressions as a group will help clarify the process. |

Rules for Order of Operations

|  |  | Do all calculations within parentheses before operations outside. |
|---|---|---|
|  | **E** | Evaluate all exponential expressions. |
|  | **MD** | Do all multiplications and divisions in order from left to right. |
|  | **AS** | Do all additions and subtractions in order from left to right. |

1.  Before you begin simplifying expressions, study the rules for order of operations above. Assign each group member to one of the steps listed. Write the name of the group member next to his or her assigned task in the table above. Note that the first step (calculations within parentheses) is not assigned. All group members will do this step together.

2.  NOTE:  If your group has done Activity 1.9, and is familiar with the group method for simplifying expressions, you may skip ahead and simplify the expression on the next page.

    Now you are ready to simplify expressions as a group. Analyze the expression together and decide on the first step. If there are parentheses, decide whether the expression inside the parentheses needs to be simplified. Following the order of operations, **E** will perform his or her task before **MD**, and **MD** will perform his or her task before **AS**.

    Practice on the example on the next page. (This is Example 4, Section 3.7 in your textbook.) The first step has been done for you: Add and subtract inside the parentheses. **AS** will do this step, writing "**AS**" in the left box, and writing the new expression below the original expression.

    Continue simplifying the expression by passing the problem to the appropriate group member for the next step. When you are done, compare your steps to those in Example 4, Section 3.7 in your textbook. If there are any discrepancies, discuss them within your group. Compare your result with the other groups. Are they the same? Discuss any differences with the other groups.

Example 4, Section 3.7

| | |
|---|---|
| | $\left(\dfrac{7}{8} - \dfrac{1}{3}\right) \times 48 + \left(13 + \dfrac{4}{5}\right)^2$ |
| **AS** | $\left(\dfrac{13}{24}\right) \times 48 + \left(\dfrac{69}{5}\right)^2$ |
| | |
| | |
| | |
| | |
| | |

3.  Once you understand the process, choose an expression from Exercises 1–18 in Exercise Set 3.7 in your textbook to simplify as a group. Use the table on the next page to organize your work. Make as many copies as you need. Alternatively, you can draw the table on a blank sheet of paper.

Do as many problems as you can in the time allotted. You can also choose one of the expressions from Exercises 29-32 in the exercise set.

| Conclusion | This activity should help you gain a better understanding of the rules for order of operations. You can also use this group method when you encounter the order of operations later on in the textbook in Sections 4.4, and 10.5. For your convenience, there are separate activities for these sections. See the table of contents. |
|---|---|

Original expression _____

| | |
|---|---|
| | |
| | |
| | |
| | |
| | |
| | |
| | |
| | |
| | |
| | |
| | |
| | |
| | |

## Activity 4.4    Use the order of operations as a group to simplify expressions.

| | |
|---|---|
| Focus | Order of operations |
| Time | 20–30 minutes |
| Group size | 3 |
| Background | Simplifying expressions using the rules for order of operations can be quite confusing for complicated expressions. Learning to simplify expressions as a group will help clarify the process. |

Rules for Order of Operations

| | | |
|---|---|---|
| | | Do all calculations within parentheses before operations outside. |
| | **E** | Evaluate all exponential expressions. |
| | **MD** | Do all multiplications and divisions in order from left to right. |
| | **AS** | Do all additions and subtractions in order from left to right. |

1. Before you begin simplifying expressions, study the rules for order of operations above. Assign each group member to one of the steps listed. Write the name of the group member next to his or her assigned task in the table above. Note that the first step (calculations within parentheses) is not assigned. All group members will do this step together.

NOTE: If your group has done Activity 1.9 or 3.7, and is familiar with the group method for simplifying expressions, you may skip ahead and simplify the expression on the next page.

2. Now you are ready to simplify expressions as a group. Analyze the expression together and decide on the first step. If there are parentheses, decide whether the expression inside the parentheses needs to be simplified. Following the order of operations, **E** will perform his or her task before **MD**, and **MD** will perform his or her task before **AS**.

Practice on the example on the next page. (This is Example 12, Section 4.4 in your textbook.) The first step has been done for you: Subtract inside the parentheses. **AS** will do this step, writing "**AS**" in the left box, and writing the new expression below the original expression. Continue simplifying the expression by passing the problem to the appropriate group member for the next step. When you are done, compare your steps to those in Example 12, Section 4.4 in your textbook. If there are any discrepancies, discuss them within your group. Compare your result with the other groups. Are they the same? Discuss any differences with the other groups.

Example 12, Section 4.4

| | $(5 - 0.06) \div 2 + 3.42 \times 0.1$ |
|---|---|
| **AS** | $4.94 \div 2 + 3.42 \times 0.1$ |
| | |
| | |
| | |
| | |
| | |
| | |

3.  Once you understand the process, choose an expression from Exercises 49–68 in Exercise Set 4.4 in your textbook to simplify as a group. Use the table on the next page to organize your work. Make as many copies as you need. Alternatively, you can draw the table on a blank sheet of paper.

Do as many problems as you can in the time allotted. Make sure you choose at least one of the more complicated expressions from Exercises 65–68.

| Conclusion | This activity should help you gain a better understanding of the rules for order of operations. You can also use this group method when you encounter the order of operations later on in the textbook in Section 10.5. For your convenience, there is a separate activity for this section. See the table of contents. |
|---|---|

Original expression _____

| | |
|---|---|
| | |
| | |
| | |
| | |
| | |
| | |
| | |
| | |
| | |
| | |
| | |
| | |
| | |

## Section 4.6     Estimate the food cost for catering a party.

| | |
|---|---|
| Focus | Decimal arithmetic |
| Time | 15 - 20 minutes |
| Group size | 3 |
| Materials | Calculator (optional) |
| Background | In this activity, each group will play the role of a catering company, and plan the menu for a party. You will use your estimation skills to prepare a food budget, and present a brief report to the rest of the class. |

1.     Each group will select one of the following catering jobs. If there are more than five groups in the class, some groups may choose the same job. Alternatively, make up a new catering job, and list it at the bottom of the table. Write the names of the group members in your group next to the chosen catering job.

| Catering Job | Group Members |
|---|---|
| Backyard social for 25 people | |
| Dessert social for 50 people | |
| Breakfast business meeting for 20 people | |
| Club luncheon for 30 people | |
| Reunion dinner for 40 people | |

2.  Now, discuss the menu for your group's catering situation. Study the catalog of items on the pages following the end of this activity, and select the specific food items that you will be serving at the party. Each group member takes turns selecting a different item. Discuss each choice, so that the overall menu is nutritionally and tastefully balanced! Write these food items in column 1 of the table below. Choose an appropriate number of items for your catering job; you do not need to use all the rows of the table. Leave the other columns blank for now.

| Food Item | Unit Price | Quantity | Estimated Cost |
|---|---|---|---|
|  |  |  |  |
|  |  |  |  |
|  |  |  |  |
|  |  |  |  |
|  |  |  |  |
|  |  |  |  |
|  |  |  |  |
|  |  |  |  |
|  |  |  |  |
|  |  |  |  |
|  |  |  |  |
|  |  |  |  |
|  |  | Total Cost |  |

3.   Next, one group member enters the unit prices in column 2 for each of the selected items. The other group members estimate the quantity of each food item needed to serve the estimated number of people attending your party. Look at the catalog list, and use the size or serving specifications to help you make this estimate. Write the required quantity of each food item in column 3 of the table above.

4.   Now you are ready to estimate the budget for your menu. Exact amounts are not needed, so use the estimation procedures in Section 4.6 of your textbook to estimate the cost of each item on your menu. Split the work, so each group member does the estimate for a few items. Write the estimated amounts in column 4 of the table on the previous page.

     When all group members are done, add the estimated costs to get the total budget for the food.

5.   Role play your group's job as caterers, and prepare a brief report suitable for presentation to a potential customer. Use the space below to write your report; be sure to discuss the menu choices, the estimated costs, and the total food budget for the party. Choose one of your group members to present the report to the rest of the class. Remember that they represent your potential customer, so be sure to put your best foot forward. After all, you want to impress them so they will hire you for the catering job!

| Conclusion | Planning a party involves arithmetic skills as well as estimation skills. This activity gives you some practical experience with the types of calculations typically associated with the catering business. |
|---|---|

| Food Item* | Size | Unit Price |
|---|---|---|
| **Beef** | | |
| Beef Patties | 12 4 - oz. patties | $7.99 |
| Chopped Beef Steaks | 16 5 1/3 - oz. patties | $14.99 |
| Beef Sandwich Steaks | 6 4 - oz. servings | $9.95 |
| Top Sirloin Steaks | 4 10 - oz. steaks | $22.69 |
| Sirloin Fillet Beef Steaks | 8 4 - oz. steaks | $14.29 |
| Salisbury Steak | 8 4 - oz. portions | $6.99 |
| Italian Style Meatballs | 2.5 lb bag | $8.99 |
| Sliced Meatloaf | 16 2 - oz. slices | $7.99 |
| **Chicken/Turkey** | | |
| Lemon Pepper Chicken | 8 portions | $13.99 |
| Turkey Breast Filets | 12 portions | $14.29 |
| Chicken Breast Meat for Fajitas | 1.75 lbs | $15.29 |
| Chicken Drummies | 3.5 lbs | $9.49 |
| Hot Wings | 2.5 lbs | $9.29 |
| Bar-B-Que Wings | 2.5 lbs | $9.49 |
| Chicken Kiev | 4 5 - oz. portions | $9.29 |
| **Pork** | | |
| Boneless Pork Chops | 8 6 - oz. chops | $25.69 |
| Summer Sausage | 2.1-lb stick | $7.69 |
| Pork Spare ribs | 2.5 lb | $11.99 |
| BBQ Port Ribs | 8 6 - oz. portions | $15.29 |
| Old-Fashioned Wieners | 20 links | $10.99 |
| Country Sausage Links | 48 1 - oz. links | $10.69 |
| Thick Sliced Bacon | 2 1.5 - lb packages | $9.95 |
| Fresh Bratwurst | 15 links | $10.99 |
| Corn Dogs | 16 corn dogs | $6.99 |
| **Seafood** | | |
| Cod Fish Nuggets | 60 nuggets | $11.99 |
| Breaded Haddock Sticks | 20 1 - oz. sticks | $7.99 |
| Catfish Fillet Fingers | 40 portions | $15.69 |
| Breaded Fantail Shrimp | 2 lbs | $21.99 |
| Unbreaded Shrimp | 2 1 - lb bags | $27.99 |
| Breaded Clam Strips | 2.5 lbs | $9.95 |
| **Pizza** | | |
| Deep Dish Pizza | 19.4 oz | $3.69 |
| Single Serve Pizza | 8 5.5 - oz. pizzas | $9.95 |

| Food Item* | Size | Unit Price |
|---|---|---|
| **Side Dishes** | | |
| Mashed Potatoes | 8 servings | $2.79 |
| Scalloped Corn | 2 1 - lb trays | $5.49 |
| Shredded Hash Browns | 12 patties | $2.49 |
| Processed Cheese | 24 slices | $10.99 |
| Pasta Primavera | 2 1 - lb trays | $5.49 |
| French Fries | 5 lbs | $4.49 |
| Breaded Onion Rings | 2 lbs | $4.49 |
| Frozen Bread Dough | 3 18 - oz. loaves | $2.59 |
| Garlic French Bread | 6 3.4 - oz. portions | $7.29 |
| Chicken Egg Rolls | 16 2 - oz. portions | $10.95 |
| Corn on the Cob | 6 count | $3.29 |
| Green Peas | 2.5-lb bag | $3.49 |
| Summer Garden Pasta Blend | 2-lb bag | $3.99 |
| **Beverages** | | |
| Tea Concentrate | 6 7.5 - fl. oz. cartons | $4.99 |
| Juice Concentrate | 6 8 - fl. oz. cartons | $7.99 |
| Gourmet Ground Coffee | 2 10 - oz. bags | $8.99 |
| Fruit Punch | 6 8 - fl. oz. cartons | $4.69 |
| **Breakfast** | | |
| Ham and Cheese Omelet | 8 3.5 - oz. servings | $7.99 |
| Cinnamon Rolls | 24 2 - oz. servings | $9.69 |
| Buttermilk Pancakes | 24 pancakes | $4.99 |
| Waffles | 32 waffles | $6.69 |
| **Dessert** | | |
| Apple Pie | 3 lb. 1 oz. | $6.69 |
| Cherry Pie | 2 lb. 15 oz. | $7.69 |
| Peach Pie | 2 lb. 15 oz. | $6.69 |
| Chocolate Chip Cookie Dough | 40 cookies | $6.49 |
| Ice Cream | ½-gallon carton | $4.79 |
| Fat-Free Ice Cream | ½-gallon carton | $4.79 |
| Frozen Yogurt | ½-gallon carton | $3.79 |
| Ice Cream Cups | 24 count | $7.99 |
| Ice Cream Sandwiches | 24 count | $7.69 |
| Sundae Cones | 12 count | $7.99 |
| Ice Cream Bars | 24 count | $7.99 |
| Fudge Sticks | 24 count | $6.49 |
| Popsicles | 24 count | $6.99 |

* Catalog adapted from Schwan's Sales Enterprises, Inc. Summer 1997 catalog of quality foods.

## Activity 5.1    Analyze the ratios of different colored M&M's candies.

| | |
|---|---|
| Focus | Ratios |
| Time | 15–20 minutes |
| Group size | 3–4 |
| Materials | Calculator, one 3-oz. package of M&M's candies per group, and one 16-oz. package of M&M's candies for every four groups |
| Background | M&M's candies come in several colors, such as red, green, and blue.  In this activity, you will analyze the color composition of a package of M&M's candies using ratios, and use these ratios to predict the composition of larger packages of M&M's candies. |
| Instructor notes | If you do not have enough class time, assign step 6 for homework. |

1.  Open the 3-oz. package of M&M's candies and count the number of candies of each color. Each group member select a color for counting.  Write the results below.  If your package has colors that are not listed, write down the colors and the quantity in the blank rows at the bottom of the table.  Add the quantity of each color, and write the total in the space indicated at the bottom.

| 3-oz. package | |
|---|---|
| **Color** | **Quantity** |
| Red | |
| Orange | |
| Yellow | |
| Blue | |
| Green | |
| Brown | |
| | |
| Total | |

2. Now, let's make some general observations on the colors of the M&M's candies.

Write down the colors in descending order of the quantity of each color.

Which color occurs most often? Which color occurs least often? Use complete sentences to write down your answers.

3. Next, work with 3 other groups to analyze a 16-oz. package of M&M's. Write your results in the table below. Again, split the work to help speed up the counting process. Separate these candies from the M&M's taken from the 3-oz. package, and don't eat any of them!

| 16-oz. package | |
|---|---|
| **Color** | **Quantity** |
| Red | |
| Orange | |
| Yellow | |
| Blue | |
| Green | |
| Brown | |
| | |
| | |
| Total | |

As before, write down the colors in descending order of the quantity of each color.

Which color occurs most often? Which color occurs least often? Be sure to use complete sentences to write your answers.

Is the order of the quantity of each color different for the 3-oz. package and the 16-oz. package?

Compare your results with that of the other groups in the class. Briefly discuss any differences, and speculate on the probable reasons for the differences.

4.  Next, we will continue the analysis of the candy colors by calculating several ratios. Split the work among the members of your group, and calculate the stated ratios as fractions in simplified form, and then as unit ratios. Write your results in the table on the next page.

    Use the methods shown in Section 5.1 of your textbook to find the ratios as fractions.

    To get the unit ratios, carry out the division indicated by the ratio. You may want to use a calculator for this step. Round all answers to the nearest thousandth.

|  | 3-oz. package | | 16-oz. package | |
| --- | --- | --- | --- | --- |
|  | **Simplified Fraction** | **Unit Ratio** | **Simplified Fraction** | **Unit Ratio** |
| Number of red candies / Total number of candies | | | | |
| Number of orange candies / Total number of candies | | | | |
| Number of yellow candies / Total number of candies | | | | |
| Number of blue candies / Total number of candies | | | | |
| Number of green candies / Total number of candies | | | | |
| Number of brown candies / Total number of candies | | | | |
| Number of red candies / Number of blue candies | | | | |
| Number of brown candies / Number of green candies | | | | |
| Number of yellow candies / Number of orange candies | | | | |

5.  Compare the unit ratios and the fraction ratios. Which type of ratio do you find easier to interpret? Why do you think this is the case? Use complete sentences in your answer.

Are the unit ratios the same for the 3-oz. and the 16-oz. packages? Speculate on why this is so. Use complete sentences in your answer.

What conclusions can be drawn from the ratio of the numbers of red to blue candies? If you bought a 6-oz. package of M&M's candies, should you expect the same red to blue ratio? Why or why not? Use complete sentences in your answer.

6.  Use the unit ratios to predict how many M&M's candies of each color you would find in jars of different sizes. The total number of M&M's candies in each jar is given in the table on the next page. Round your answers to the nearest whole number, and write your results in the table.

|         | Jar 1 | Jar 2 | Jar 3  |
|---------|-------|-------|--------|
| Total   | 100   | 1000  | 10,000 |
| Red     |       |       |        |
| Orange  |       |       |        |
| Yellow  |       |       |        |
| Blue    |       |       |        |
| Green   |       |       |        |
| Brown   |       |       |        |

7.  Now that you have concluded the exercises in this activity, take a break and eat the M&M's candies!

| Conclusion | The analysis of M&M's candies provides an interesting and tasty application of ratios.  You should note that the unit ratios in this activity are very similar to the unit rates discussed in Section 5.2 of your textbook.  Also, the method shown in step 6 for predicting the quantity of each color is very similar to the method discussed in Section 5.3 for solving proportions. |
|------------|---|

## Activity 5.4    Use proportions to make predictions of your college's student population.

| | |
|---|---|
| Focus | Applications of Proportions |
| Time | 20–30 minutes |
| Group size | 3–4 |
| Materials | Calculator (optional) |
| Background | One of the many applications of proportions is in the field of population demographics. Statisticians use the characteristics of a small population sample to predict the characteristics of the entire population. For an accurate prediction, the sample should be unbiased, and represent a typical cross-section of the population. To conduct the study, a poll is taken of the sample population, and the results extrapolated to the entire population. In this activity, your class will be the sample population. You will conduct a poll of the students in the class, and use the results to predict the demographics of your college's student population. |

1.  The first step in this study is to take a poll of all the students in your class, using the table on the next page. You will do this together with the other groups. Select three individuals from the entire class to conduct the poll. One student writes the results on the board, while the other two count the number of students in each category. If the class would like to collect data on a characteristic not listed, use the blank rows at the bottom of the table to record this information.

Within each group, one group member should record the data in the table on the next page.

Once you have finished the poll, there are two more pieces of data you will need before you can begin the analysis. Count the total number of students in you class, and also find out how many students are enrolled in your college. You may need to estimate this number. Write these numbers below the table on the next page.

| Class Population | |
|---|---|
| **Characteristic** | **Number of students** |
| Male | |
| Female | |
| Single | |
| Married | |
| Right-handed | |
| Left-handed (or ambidextrous) | |
| Full-time student | |
| Part-time student | |
| Full-time job | |
| Part-time job | |
| | |
| | |
| | |
| | |
| | |

Total Class Population  _____

Total College Population  _____

2. Now, we will use proportions to make predictions about the college student population. For each characteristic, translate into a proportion as shown here:

$$\frac{\text{\# of Students in Class with Characteristic}}{\text{Total Class Population}} = \frac{\text{\# of Students in College with Characteristic}}{\text{Total College Population}}$$

Solve each proportion for the unknown value, and round all answers to the nearest whole number. Split the work among all group members, and use a calculator to help speed up your calculations. Record your results in the table below.

| College Population | |
|---|---|
| **Characteristic** | **Number of students** |
| Male | |
| Female | |
| Single | |
| Married | |
| Right-handed | |
| Left-handed (or ambidextrous) | |
| Full-time student | |
| Part-time student | |
| Full-time job | |
| Part-time job | |
| | |
| | |
| | |
| | |
| | |

3. Before going any further, look at the results in the table above. Do the numbers look accurate? Check the arithmetic of any numbers that you have doubts about. Once all group members are satisfied with the accuracy of the predictions, proceed to the next step.

4. Now, analyze each prediction of the college student population. Discuss in your group how reasonable each prediction looks. Base your arguments on your observations on the student population and the likelihood of each prediction. Use complete sentences and write down a summary of your group's discussion below. List any prediction you believe to be inaccurate in the first column of the table below.

| Inaccurate Predictions | Possible Reasons for Inaccuracy | Possible Remedies |
|---|---|---|
|  |  |  |
|  |  |  |
|  |  |  |
|  |  |  |
|  |  |  |
|  |  |  |
|  |  |  |

5.  Consider the predictions from step 4 that your group highlighted as inaccurate. For each of these predictions, discuss possible reasons for the inaccuracy. Write these down in the second column of the table on the previous page. Then, discuss possible remedies to improve accuracy and write these down in the third column.

6.  Each group should present a brief report to the entire class. Discuss the predictions your group labeled as inaccurate, the possible reasons for the inaccuracy, and the possible remedies.

| Conclusion | As you observed, the study of population demographics is an inexact science. While the mathematics is a fairly straightforward application of proportions, the reasonableness of the predictions depends on many factors. Some factors can be controlled by choosing a representative sample population, but there usually remains some amount of uncertainty. Think about this next time you read about population demographics in the newspapers or magazines. The numbers reported are usually based on a poll of a sample population, and may be grossly inaccurate! |

Name _____  Section _____  Date _____

## Activity 6.2　Use percent squares to develop a number sense for percents.

| | |
|---|---|
| Focus | Percents and fractions |
| Time | 15–20 minutes |
| Group size | 4 |
| Background | In Section 6.1 of your textbook, the percent square was used to show the meaning of percent. In the example given, 70% is a ratio of 70 to 100, and 70 of 100 squares are shaded. In this activity, percent squares are used to help you develop a number sense for percents. |

1.　Each group member chooses two different percent values between 1% and 99%. Write your choices below, but do not show them to the other group members.

Shade these percents on the two percent squares below. You can shade the squares in any pattern you like, but <u>do not</u> write down the percent values. The next part of this exercise will require you to guess the percent values from another group member's paper.

Give the shaded percent squares to the group member to your left. On the paper you receive, guess the percent value represented by each shading. Do not count squares!

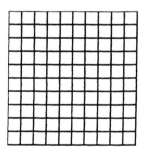
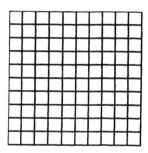

Your guess _____ %　　　　Your guess _____ %

Actual value _____ %　　　　Actual value _____ %

When you are done, return the paper to the original group member. The actual values should now be written down below each percent square. If the values differ greatly, discuss this with the appropriate group member and make necessary changes.

2. We will now look at fraction squares and see how they relate to percent squares. Study the fraction squares below. As a group, guess the percent values for each square, and write your guesses below the appropriate square.

Next, use the method shown in Section 6.2 of your textbook to convert each fraction to a percent. Split the work up so that each group member only has to convert one fraction.

How well do your guesses compare with the actual percent values? Write down some strategies that can help you improve the accuracy of your guesses.

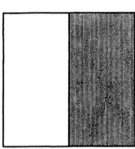

Fraction $\dfrac{1}{2}$

Percent _____%

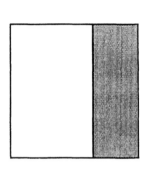

Fraction $\dfrac{3}{8}$

Percent _____%

Fraction $\dfrac{1}{3}$

Percent _____%

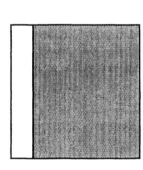

Fraction $\dfrac{5}{6}$

Percent _____%

4. Think about the strategies you used to guess the percent values of the fraction squares in step 3. Would this strategy work when you are given only the fraction name?

Let's practice by playing the following game. Choose three fractions that are smaller than 1. Write them below, but do not let the other group members see your paper.

| Fraction | Percent |
|----------|---------|
|          |         |
|          |         |
|          |         |

Convert each fraction to a percent, using the method shown in Section 6.2 of your textbook. Now, you are ready to begin the game.

One group member starts by writing his or her first fraction on a blank sheet of paper. The other group members have 10 seconds to guess the percent value for that fraction. Remember that the goal is to develop a number sense for percents, so pencil and paper conversions are not allowed at this step. The group member with the closest guess gets one point.

Continue with the next group member, until each one has presented all the fractions. Skip any repeated fractions. The group member with the most points at the end is the winner, and has a good number sense for percents!

| Conclusion | The exercises in this activity should give you a better number sense for percents. Continue developing this sense by mentally visualizing the percent square anytime you encounter percents or fractions. You can also use this technique when converting mixed numerals to percents. |
|------------|------|

## Activity 6.6    Calculate the costs associated with the purchase of a car or truck.

| | |
|---|---|
| Focus | Sales tax, commission, discount |
| Time | 20–30 minutes |
| Group size | 2 |
| Materials | Calculator, photos and ads for your dream car or truck (optional) |
| Background | Purchasing a car or truck can be a traumatic experience.  In addition to the negotiations between the buyer and salesman, there are several mathematical calculations that need to be performed.  As a buyer, you need figure out what purchase price you can afford based on your monthly budget.  As a car salesman, you need to be able to figure out your commission, and any discounts offered to the customer. |
| Instructor notes | You can assign roles in one class period, and have the tables in steps 2 and 3 done outside class as homework.  Then, continue with the rest of the activity in the next class period. |

1.    Choose the role of the buyer or the salesman, and write your names in the appropriate space below.

Buyer _____

Salesman _____

The buyer will be visiting the car dealership to buy a car or truck from the salesman.  The buyer's goal is to negotiate the lowest price on the desired vehicle, while keeping within his or her budget.  The salesman, on the other hand, will want to negotiate the highest price possible, since he or she is paid on commission.

2.  (If you are the salesman, skip to step 3.)

Most buyers use the monthly payment to help decide how much they can spend on a car or truck. Fill out the table on the next page, and use it as a reference during your negotiations with the salesman. Keep this table secret from the salesman! These formulas will help you with your calculations.

$$\text{Sales Tax} = \text{Sales Tax Rate} \cdot \text{Purchase Price}$$

(Use your state sales tax rate for automobiles to figure out the sales tax.)

$$\text{Loan Amount} = \text{Purchase Price} + \text{Sales Tax}$$

$$\text{Monthly payment} = (0.0222)\,(\text{Loan Amount})$$

(The monthly payment is based on a 5 year loan at 12% interest. You can also use the information in the activity for section 6.7 to calculate monthly payments for other interest rates and years.)

3.  (If you are the buyer, skip to step 4.)

The salesman should have a rough idea of the commissions he or she will receive from the sale. In addition, your manager has authorized you to offer a discount on the purchase price. Fill out the discount and commission table on the next page, and use it as a reference during your negotiations with the buyer. Keep this table secret from the buyer! These formulas will help you with your calculations.

$$\text{Discount} = \text{Discount Rate} \cdot \text{Purchase Price}$$

(The discount rate is typically set by the manager of the dealership. For this activity, however, choose a discount rate between 1% and 10%.)

$$\text{Discounted price} = \text{Purchase Price} - \text{Discount}$$

$$\text{Commission} = \text{Commission Rate} \cdot \text{Purchase Price}$$

(Your commission is based on the full purchase price. As with the discount rate, the commission rate is determined by the dealership you work for. However, for this activity, choose a rate between 1% and 10%.)

# BUYER'S REFERENCE TABLE

| Purchase Price | Sales Tax | Loan Amount | Monthly Payment |
|----------------|-----------|-------------|-----------------|
| $10,000 | | | |
| $11,000 | | | |
| $12,000 | | | |
| $13,000 | | | |
| $14,000 | | | |
| $15,000 | | | |
| $16,000 | | | |
| $17,000 | | | |
| $18,000 | | | |
| $19,000 | | | |
| $20,000 | | | |

# SALESMAN'S REFERENCE TABLE

| Purchase Price | Discount | Discounted Price | Commission (use the purchase price!) |
|---|---|---|---|
| $10,000 | | | |
| $11,000 | | | |
| $12,000 | | | |
| $13,000 | | | |
| $14,000 | | | |
| $15,000 | | | |
| $16,000 | | | |
| $17,000 | | | |
| $18,000 | | | |
| $19,000 | | | |
| $20,000 | | | |

4. Both of you are now ready to meet. Role play your characters, and talk about the make, model, color, and price of several cars or trucks. The buyer needs to weigh the features of the car or truck against the price he or she is willing to pay. Consult your monthly payment table and try to reach a compromise with the salesman. You can also use the photos and ads to support your position.

The salesman, on the other hand, wants to earn the highest possible commission, but cannot afford to scare away the buyer. Negotiate as best as you can, and don't be afraid to offer a discount to the buyer if you need to close the deal. You can offer any dollar amount of discount up to the discounted price authorized by your manager, but you want to try and get the highest price for the dealership. Once you have settled on a price, write your result below.

| BUYER | | SALESMAN | |
|---|---|---|---|
| Purchase Price | | Purchase Price | |
| Loan Amount | | Discounted Price | |
| Monthly Payment | | Commission | |

| Conclusion | This activity gives you a practical application of percents. Use the techniques shown to help you budget for a new car or truck. You can also continue with the next activity (Section 6.7) to calculate a payment schedule for the car loan. |
|---|---|

## Activity 6.7    Prepare an amortization table for a car loan.

| | |
|---|---|
| Focus | Simple and compound interest |
| Time | 20–30 minutes |
| Group size | 2 |
| Materials | Calculator |
| Background | When you take out a loan to purchase a car or truck, the loan is usually paid off in monthly installments. The payment amount is based on the principal, interest rate, and length of the loan. We say that the loan is *amortized* over the loan period, and an *amortization schedule* can be prepared to show you how much principal is still owing at the end of each month. |

1. Study the amortization schedule below. This is for a loan of $10,000 for one year at 12% interest compounded monthly.

| Month | Payment Amount, $M$ | Interest, $I$ | Principal Remaining, $P$ |
|:---:|:---:|:---:|:---:|
| 0 | - | - | $10,000.00 |
| 1 | $888.49 | $100.00 | $9,211.51 |
| 2 | $888.49 | $92.12 | $8,415.14 |
| 3 | $888.49 | $84.15 | $7,610.80 |
| 4 | $888.49 | $76.11 | $6,798.41 |
| 5 | $888.49 | $67.98 | $5,977.91 |
| 6 | $888.49 | $59.78 | $5,149.20 |
| 7 | $888.49 | $51.49 | $4,312.20 |
| 8 | $888.49 | $43.12 | $3,466.83 |
| 9 | $888.49 | $34.67 | $2,613.01 |
| 10 | $888.49 | $26.13 | $1,750.65 |
| 11 | $888.49 | $17.51 | $879.67 |
| 12 | $888.49 | $8.80 | -$0.01 |

The payment amount is calculated using the formula

$$M = \frac{rA(1+r)^n}{(1+r)^n - 1},$$

where  $M$  is the monthly payment
$r$  is the interest rate for each compounding period
$A$  is the amount of money loaned
$n$  is the number of compounding periods.

For the example given,   $r = 12\% \div 12 = 1\%$ or $0.01$
$n = 1$ year x $12$ months $= 12$
$A = \$10,000$.

Substituting these values into the formula, we get

$$M = \frac{(0.01)(10,000)(1.01)^{12}}{(1.01)^{12} - 1} = \$888.49.$$

The interest owed each month is calculated using the simple interest formula, $I = P \cdot r \cdot t$. This is the same formula given in Section 6.7 of your textbook, and $P$ is the principal remaining, $r$ is the monthly interest rate, and $t$ is the time period.

Thus, for the first month,        $I = P \cdot r \cdot t = (10{,}000)(0.01)(1) = \$100$.

The remaining principal is found by subtracting the monthly payment from the previous month's principal, and adding the interest owed for that month.

For the first month,                $P = 10{,}000 - 888.49 + 100.00 = 9211.51$.

Notice that the principal remaining of $\$0.01$ at the end of month 12 should be ignored, as it is due to round-off error.

2.   Once you understand how the table is set up, you are ready to create your own amortization schedule.  First, decide as a group what the terms of the loan will be.  Write down the information below.

| | |
|---|---|
| Amount of loan, $A$ | |
| Annual interest rate | |
| Monthly interest rate, $r$ | |
| Number of years | |
| Number of months, $n$ | |

3. Now, use the formula from Step 1 to calculate the monthly payment, *M*. Show your work below.

4. Next, fill out the amortization schedule on the next page. One group member should do the calculations on a calculator, while the other group member records the results on the table. If your loan is for a longer period than 24 months, you can continue the schedule on another sheet of paper.

5. When you have finished filling out the amortization schedule, answer the following questions.

   For the first month, what percent of the monthly payment goes towards paying the interest?

   What is the total of all monthly payments?

   What is the total interest paid on this loan? (Be careful! There are two ways to calculate the total interest. One way is very long!)

   What percent of the total of all monthly payments goes towards paying the total interest?

   What effect does decreasing the interest rate have on the total interest and monthly payment? Use complete sentences in your answer.

   What effect does increasing the interest rate have on the total interest and monthly payment?

| Month | Payment Amount, $M$ | Interest, $I$ | Principal Remaining, $P$ |
|:-----:|:-------------------:|:-------------:|:------------------------:|
| 0     | -                   | -             |                          |
| 1     |                     |               |                          |
| 2     |                     |               |                          |
| 3     |                     |               |                          |
| 4     |                     |               |                          |
| 5     |                     |               |                          |
| 6     |                     |               |                          |
| 7     |                     |               |                          |
| 8     |                     |               |                          |
| 9     |                     |               |                          |
| 10    |                     |               |                          |
| 11    |                     |               |                          |
| 12    |                     |               |                          |
| 13    |                     |               |                          |
| 14    |                     |               |                          |
| 15    |                     |               |                          |
| 16    |                     |               |                          |
| 17    |                     |               |                          |
| 18    |                     |               |                          |
| 19    |                     |               |                          |
| 20    |                     |               |                          |
| 21    |                     |               |                          |
| 22    |                     |               |                          |
| 23    |                     |               |                          |
| 24    |                     |               |                          |

6. You can also use the monthly payment formula to figure out the loan amount for a given monthly payment. The version of the formula you will need is

$$A = \frac{M\left[(1+r)^n - 1\right]}{r(1+r)^n}.$$

where    $M$   is the monthly payment
            $r$   is the interest rate for each compounding period
            $A$   is the amount of money loaned
            $n$   is the number of compounding periods

Suppose you budget $300 a month for a new car loan. You do some research and find that you can get a loan for 60 months at 12% annual interest. Using the formula, we can find the maximum purchase price you can afford. For this example,

$$M \;=\; \$300.00$$
$$r \;=\; 12\% \div 12 \;=\; 1\% \text{ or } 0.01$$
$$n \;=\; 60.$$

Substituting these values into the formula, we get

$$A \;=\; \frac{300.00\left[(1.01)^{60} - 1\right]}{(0.01)(1.01)^{60}} \;=\; \$13{,}486.51.$$

Thus, you can afford to buy a car that costs $13,486.51 or less. Now, assume that one of you is planning to purchase a new car. Set a budget for the monthly loan payment, and choose reasonable values for the interest rate and loan period. Use these values to calculate the maximum price of the car you can afford. Show your work below.

| Conclusion | Buying a car is an event you are likely to encounter in the future. This activity shows you how to calculate your monthly payments and figure out the amortization schedule for the car loan. You can also use the methods shown to prepare an amortization schedule for other kinds of loans. If you have access to a computer, you can use a spreadsheet program to create an amortization schedule without having to do all the calculations by hand. The activity in the previous section, Section 6.6, gives you more detail on the mathematics involved with purchasing a car. If you haven't already done that activity, you may want to try it. |
|---|---|

## Activity 7.1    Perform a statistical analysis of pulse rates.

| Focus | Averages, medians, and modes |
|---|---|
| Time | 15–20 minutes |
| Group size | 5 |
| Materials | Stopwatch, calculator |
| Background | The definitions of average, median, and mode are given in Section 7.1 of your textbook. In this activity, you will gather data on pulse rates and analyze the data by calculating the average, median, and mode. |
| Instructor notes | Be prepared to suggest alternative activities for students with physical disabilities. |

1.    First, locate your pulse. There are two places you can most easily find it. The first, called the **radial** pulse, is located on your wrist. Turn your palm up, and place the forefinger of your other hand on your wrist below the base of your thumb. The second pulse, called the **carotid** pulse, is on the side of your throat below the jaw bone under your ear. Again, use your forefinger to locate this pulse.

   Once you have found your pulse, decide which one is easier to find. Then use this location for the remainder of this activity.

2.    Each group member will count his or her pulse rate under three different conditions - resting, exercise, and recovery.

   Rest for about 2 minutes, then find your pulse and count the number of beats in a 15 second period. Multiply this number by 4 to get your resting pulse rate in beats per minute. Record this number in the table on the next page. Use one table for all the members of your group.

3.    Next, measure your exercise pulse rate. Run in place for 1 minute, then immediately count your pulse for a 15 second period. Multiply this number by 4 to get your exercise pulse rate. Record this number in the appropriate column in the table on the next page.

4.    Finally, measure your recovery pulse rate. Rest for about 3 minutes, then measure your pulse rate as you did the previous two times. Record this number in the table as well.

   After all group members have recorded their pulse rates in the table, each group should write their pulse rate data on the board, following the same format from the table.

| Name | Resting Pulse | Exercise Pulse | Recovery Pulse |
|---|---|---|---|
| | | | |
| | | | |
| | | | |
| | | | |
| | | | |
| AVERAGE | | | |
| MEDIAN | | | |
| MODE | | | |

5. Now, each group will analyze their group's pulse rate data. Calculate the average, median, and mode for each of the three different pulse rates. Write your results in the appropriate spaces in the table above. Then calculate the average, median, and mode for the class pulse rate, using the data on the board.

6. How does your group's average compare with the class average? Use complete sentences in your answer.

Which average do you think is more representative of the population as a whole? Why do you believe this is so?

6. Compare the median and mode for your group's pulse rates with the class median and mode. Is there a large difference between the two sets of numbers? Why do you think this is so? Be sure to use complete sentences in your answer.

Look up the definitions of median and mode in Section 7.1 of your textbook. Do the definitions explain the differences in the results? How?

| Conclusion | This activity gives you practice in analyzing data using average, median, and mode. As you probably discovered, the larger the data set, the more representative of the population are the results of the analysis. Small data sets may give you an inaccurate picture. |
|---|---|

## Activity 7.3    Analyze class grades using line graphs and bar graphs.

| | |
|---|---|
| Focus | Bar graphs, line graphs |
| Time | 20–25 minutes |
| Group size | 4–5 |
| Materials | Ruler, calculator |
| Background | Data can be more easily visualized with graphs. The type of graph drawn is determined by the nature of the data and the analysis to be done of the data. For example, when analyzing class test grades, you would use a bar graph to show the number of students earning each grade. A line graph would be too cluttered for this purpose. On the other hand, when you want to track the grade history for one student over time, a line graph is more valuable. This activity will give you practice with both types of graphs. |
| Instructor notes | Make one transparency of the grid in step 6 per group, so the results from each group can be shown to the entire class on the overhead projector. |

1.   First, the class should choose up to 5 tests and/or quizzes for analysis. Each group member then writes down his or her grades in the table below. If necessary, convert each grade to a percent.

| Test or Quiz | Grade earned | Grade as a % |
|---|---|---|
| | | |
| | | |
| | | |
| | | |
| | | |

2.   Next, each group member draws a line graph to show how his or her grades have changed during the term. Using the grid on the next page; indicate the different tests on the horizontal scale, and mark the vertical scale appropriately to show the grade earned. Choose titles for the vertical and horizontal scales and also give an overall title to the graph.

Plot your grades on the graph, and draw line segments connecting the points.

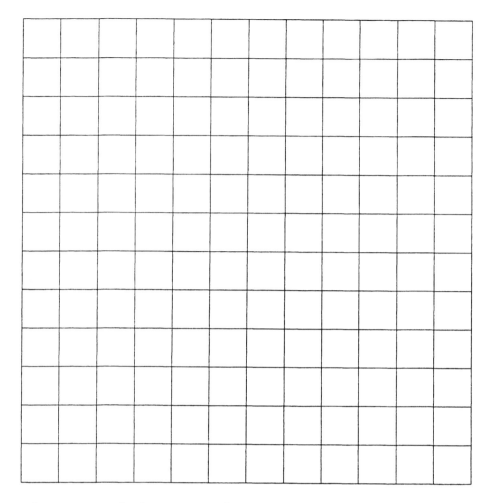

3.     Observe how your grade changes over time. Does your graph fit into one of the categories below?

                        Increasing from left to right

                        Decreasing from left to right

                        Constant from left to right

                        Irregular from left to right

What can you say about your grade performance in this class based on the shape of your graph?

If you were to take another test, what do you think your grade might be?

Exchange graphs with another group member. Analyze the graph and make some observations about his or her grade performance in the class.

4. For the second part of this activity, work as a group to analyze the class performance on one test or quiz. First, one representative from each group writes the grades for the entire group on the board. Organize the data in columns, with one test or quiz per column. The grades do not have to be in any order. When all groups are done, there should be up to 5 columns of grades on the board.

Each group chooses a different test for analysis. Write the grades for your chosen test in the table below.

| Class grades for Test _____ |
| --- |
| |

5. Begin your analysis by clustering the grades. Decide on the range of grades that correspond to each of the letter grades A, B, C, D, or F. Your instructor may have a predetermined set of values that you should use. Use the table below to record the grade ranges.

Then, count the number of students who scored in each category, and write the result in the table.

| Category | Grade Range | Number of students |
| --- | --- | --- |
| A | | |
| B | | |
| C | | |
| D | | |
| F | | |

6.  Next, use the graph below to make a vertical bar graph of the data. Indicate on the horizontal scale in five equally spaced intervals the letter grades A to F. For the number of students, choose an appropriate scale so the vertical bars are not too high or too low. Choose titles for both scales and also choose an overall title for the graph. The graph you have constructed is called a **histogram.**

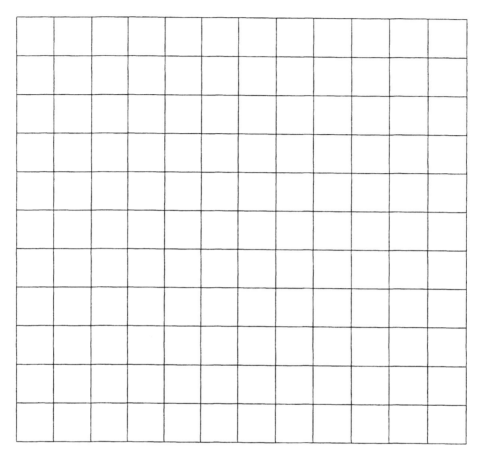

7.  Looking at the bar graph you just drew, what observations can you make about the class' grade performance on this test? Use complete sentences in your answer.

Exchange graphs with a group that analyzed a different test. What observations can you make about the class' grade performance on this test?

| Conclusion | As this activity illustrates, line graphs and bar graphs are used for different purposes. A line graph is often used to show a change over time, as was done with the grades earned during the term. A bar graph, on the other hand, is convenient for showing comparisons. When the class grades were displayed with a bar graph, it was easy to tell at a glance which grade was earned by the most students. Other similar observations were also easy to make. |
|---|---|

## Activity 7.4    Use a circle graph to show household expenses.

| | |
|---|---|
| Focus | Circle graphs |
| Time | 15–20 minutes |
| Group size | 2 |
| Materials | Ruler, calculator, color pencils |
| Background | Circle graphs are often used to show the percent of a quantity used in different categories. In this activity, you will draw a circle graph to show the household expenses for food, entertainment, and other categories |

1.  First, estimate the annual expenses of one of your group members. Round all values to the nearest hundred, and write your results in the table below. You may use the categories given, and add or modify them as needed.

| Expense | $ amount | % of total |
|---|---|---|
| Housing | | |
| Food | | |
| Transportation | | |
| Clothing | | |
| Medical care | | |
| Entertainment | | |
| Education | | |
| Other | | |
| | | |
| | | |
| **TOTAL** | | |

2.  Add the dollar amounts in the second column and write the total in the space provided. Now, use the total amount to calculate the percent spent in each category. Split the work so that each group member does the calculations for half the categories. Round your answers to the nearest percent, and write the results in the third column of the table. Check your work by adding the percents in the third column. The total should be 100%. (Note that round-off error may give you an answer that is not exactly 100%.)

3.   Draw a circle graph to show the expenses of your group. Use the circle below marked with 100 equally spaced tick marks. One group member starts by choosing one expense category. Draw a line from the center to one tick mark. Count off the appropriate number of tick marks and draw another line. Shade the wedge with one color and label it. The next group member continues by choosing another category, and drawing the wedge next to the wedge that is already drawn. You can choose the categories in any order. Continue in this way until all categories are drawn. Use a different color for each category. (If you do not have color pencils, shade each wedge with a different pattern.) Finally, give your finished graph an overall title.

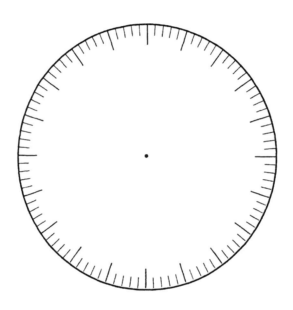

4.   Now, look at the completed circle graph and compare it to the table of expenses. The graph gives a much clearer picture of where your money is spent.

Exchange graphs with another group, and analyze their expenses. How does their spending pattern compare with yours? Do you see situations when you might want to alter your spending patterns? Use complete sentences in your answer.

| Conclusion | Circle graphs are very effective to show visually the percent of a household income spent in various categories. By comparing the size of each slice, you can easily see where you spend the most money or how the expenditures compare to each other. |
|---|---|

## Activity 8.1   Practice conversions with the old British monetary units.

| | |
|---|---|
| Focus | Unit conversions |
| Time | 10–15 minutes |
| Group size | 2 |
| Background | The old British monetary system used units such as farthings, pence, shillings, and crowns.  The conversions factors were not simply powers of 10, as they are in the current system, so this old system gives you an opportunity to practice unit conversions using a different system from the American units in your textbook. |

### Old British Monetary Units

| | | |
|---:|:---:|:---|
| 4 farthings | = | 1 pence |
| 12 pence | = | 1 shilling |
| 1 shilling | = | 1 bob |
| 5 shillings | = | 1 crown |
| 20 shillings | = | 1 pound |
| 21 shillings | = | 1 guinea |
| 1 pound | = | 1 sovereign |

1. Study the old British monetary units in the table above.  Practice some unit conversions by completing the statements below.  Work with your partner until each of you is proficient with the conversions.

5 bob   =   _____ pence

2 pounds   =   _____ bob

1 guinea   =   _____ pounds _____ shillings

1 sovereign   =   _____ shillings

Half a crown   =   _____ shillings _____ pence

Half a crown   =   _____ pence

2. Next, one group member plays the role of a shopkeeper in an old English general store, while the other group member is the customer.

   The customer should choose an item from the list below, and "pay" for the item using coins of one kind only.

   The shopkeeper will need to calculate the correct change to return to the customer.

   After the customer does this a few times, switch roles, so each group member gets practice in making change.

| Item | Cost |
| --- | --- |
| Soap | 5 pence |
| Toothpaste | 3 pence 2 farthings |
| Towel | 8 pence 1 farthing |
| Brush | 2 bob 5 pence |
| Broom | 5 bob |
| Pot | Half a crown |
| Kettle | 7 shillings 10 pence |
| Silverware | 4 shillings 6 pence |
| Dinnerware | 12 shillings 6 pence |
| Wine glasses | 10 shillings 2 pence |

3. For a more advanced exercise, the customer may choose several items and calculate the total amount of the purchases. The total should be expressed using the least number of coins. As before, the customer pays for the items with one coin, and the shopkeeper has to figure out the correct change.

| Conclusion | Even though the old British monetary system is not in use any longer, it gives us excellent practice with unit conversions. The current British system uses a decimal conversion of 100 pence = 1 pound, and is not as interesting or useful for practicing unit conversions. |
| --- | --- |

## Activity 8.6    Convert between American units and metric units of temperature, length, volume, and weight.

| | |
|---|---|
| Focus | Converting units of temperature, length, volume, and weight |
| Time | 15 –20 minutes |
| Group size | 2 |
| Materials | Calculator |
| Background | The United States is one of the few countries in the world that is not using metric units for everyday applications. Our northern neighbor, Canada, has been using metric units for about 20 years. If you should visit Canada, you will need to convert metric units to American units, at least approximately. Conversely, a Canadian resident visiting the U.S. would need to know how to convert American units to metric units. |

1.    First, as a group, complete the conversion table below. The information can be found in Sections 8.3 and 8.6 of your textbook.

| **Metric to American units** | **American to metric units** |
|---|---|
| 1 m  =  _____ ft | 1 ft  =  0.305 m |
| 1 km  =  _____ mi | 1 mi  =  _____ km |
| 1 kg  =  2.2 lb | 1 lb  =  0.454 kg |
| 1 L  =  3.78 gal | 1 gal  =  0.264 L |
| Celsius to Fahrenheit formula: | Fahrenheit to Celsius formula: |
| F  =  _____ | C  =  _____ |

2. One group member takes the role of the Canadian resident, while the other group member is the American visitor. Role play the following scenes as the American gets acquainted with the metric system in Canada.

American: Before leaving home, you call your Canadian friend to find out what the weather is like.

Canadian: The temperature is currently 23°C, and tonight the lows are around 10°C. Convert these temperatures to Fahrenheit so your American friend will know whether to pack a sweater.

American: You arrive in Canada, and your friend drives you to his or her home. You notice that the speed limit is 100 km/h, and it is 75 km to your destination. This seems awfully fast to you.

Canadian: Convert the given speed and distance to American units. Also complete the following table, so your friend can develop a feel for the metric speed limits and distances.

| km/h | mph | km | mi |
|------|-----|-----|-----|
| 110 | | 500 | |
| 100 | | 400 | |
| 90 | | 200 | |
| 80 | | 100 | |
| 70 | | 50 | |
| 60 | | 10 | |

American: On the way home, your friend stops to buy gas. The cost is 61.9 ¢/L, and you are surprised that gas is so inexpensive.

Canadian: Bring your friend down to reality (gently) by converting the gas price to cents per gallon.

American: The next morning, you are reading the newspaper and the sports page gives the statistics of a local athlete as:

Height: 1.85 m            Weight: 92 kg

Canadian: Help your friend relate to these statistics by converting them to American units.

3. It's time for a return visit as the Canadian travels to the United States to visit the American. Keep the same roles as before. Here's what the American has to do to ensure the Canadian friend adjusts to the American units.

   • Convert the current temperature to Celsius degrees.

   • Tell your Canadian friend what the normal high and low temperatures are in Celsius degrees, and give him or her a general idea of local weather conditions at this time of year.

   • Convert the speed limits of local roads and highways to km/h.

   • Suppose your friend is driving a vehicle equipped with a speedometer marked in miles per hour only. Develop a quick method for converting speed limits to km/h so he or she won't accidentally exceed the speed limit when driving in America.

   • Convert the distances to several nearby cities or towns to kilometers.

   • Convert the current local price of gas to cents per liter. Would your friend consider these prices high or low?

   • Convert your height to meters, and your weight to kilograms.

   • Discuss the height and weight of any well-known sports figure, giving the height in meters, and the weight in kilograms.

| Conclusion | If you travel to other countries in the world, you will need to understand the metric units of temperature, length, volume, and weight. It is advantageous to develop a feel for these units, and be proficient in unit conversions, so you know that 35°C is quite hot, and 100 km/h is not as fast as the number suggests! |
|---|---|

## Activity 8.7    Verify the conversions between American units of area.

| | |
|---|---|
| Focus | Converting units of area |
| Time | 10–15 minutes |
| Group size | 2 |
| Materials | Ruler |
| Background | The conversion table for American units of area is given in Section 8.7 of your textbook. This activity shows you how to verify two of the conversions, and will help you develop a numerical sense for converting units of area. |

1.  Complete the conversion table below. The values needed are given in Sections 8.1 and 8.7 of your textbook. One group member can look up the values, while the other group member records them below.

$$1 \text{ foot (ft)} = \underline{\hspace{1cm}} \text{ inches (in.)}$$

$$1 \text{ square foot (ft}^2) = \underline{\hspace{1cm}} \text{ square inches (in}^2)$$

$$1 \text{ yard (yd)} = \underline{\hspace{1cm}} \text{ feet (ft)}$$

$$1 \text{ square yard (yd}^2) = \underline{\hspace{1cm}} \text{ square feet (ft}^2)$$

2.  Using the grid on the next page, one group member draws a square that is 1 foot by 1 foot. Assume each grid square is 1 inch long, and draw your square such that each side has an appropriate number of inches to make up 1 foot.

    The other group member now counts the number of grid squares in the 1 foot square. Write this number below.

    Number of grid squares in a foot square = _____

    Since one grid square is equal to one inch square, complete the sentence below.

    _____ square inches = 1 square foot

    Compare this sentence with the corresponding sentence in your conversion table. Are they the same? Why do you think this is so? Use complete sentences in your answer.

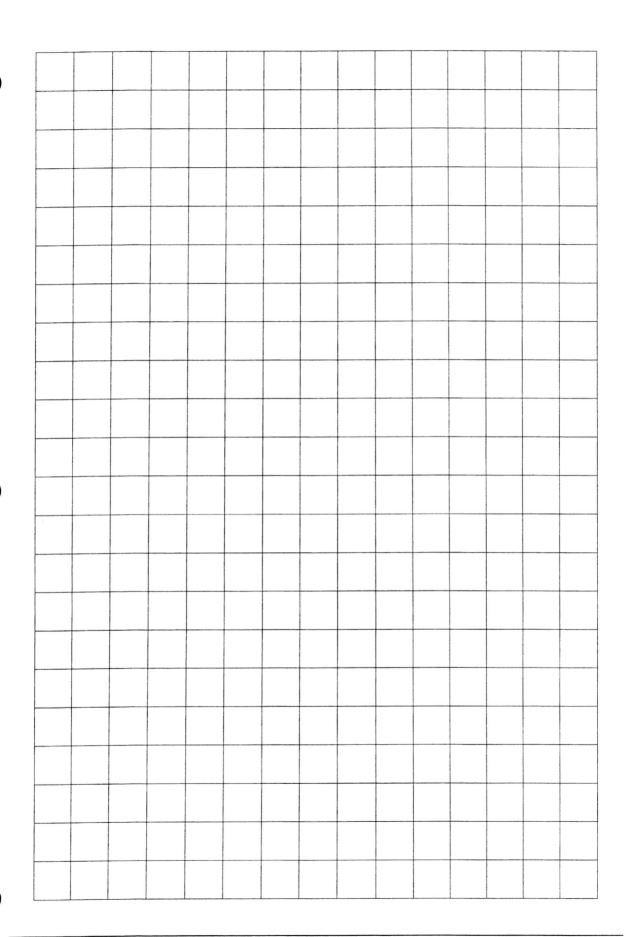

3.  Following the same procedure, draw a square that is 1 yard by 1 yard. In this case, let each grid square count as 1 foot. Switch roles, so both group members get to practice drawing and counting the squares. As before, count the number of grid squares in the 1 yard square, and complete the sentences below.

Number of grid squares in a yard square = _____

_____ square feet = 1 square yard

Again, how does this last sentence compare with the corresponding sentence in your conversion table? Be sure to use complete sentences in your answer.

4.  Now, let's generalize this concept of unit conversions for area. Each group member should create a fictional unit of length, and make it equal to an arbitrary number of some other fictional unit. Keep this information secret from your partner!

For example, you could define 1 click = 5 ticks. You could also use the first names of each group member for the unit names. Be creative!

1 _____  =  _____ _____

Following the process from steps 2 and 3, draw a square with dimensions of 1 by 1 of your large units. Each side of the square should have the appropriate of smaller units. Write the unit names next to the square, indicating which is the larger unit, but do not write the conversion sentence. Use your square to help you complete the following sentence for your fictional units.

1 square _____  =  _____ square _____

Next, exchange the drawing with your partner. Study the square you are given, and try to determine what the conversion sentence is. Check your answer with your partner, and resolve any discrepancies.

| Conclusion | In this activity, you saw that the conversions between units of area are tied to the formula for the area of a square. Knowing this makes it easy to convert from one unit of area to another when you have the conversion for one length unit to the other. |

● **Activity 9.2     Prepare a budget for redecorating the classroom.**

| | |
|---|---|
| Focus | Perimeter, area |
| Time | 15–20 minutes |
| Group size | 4 |
| Materials | Several tape measures, calculator |
| Background | The geometric concepts of perimeter and area are essential when you want to redecorate a room. For example, the perimeter of the floor is needed to estimate the cost of replacing the base trim, and the area of the floor is used to calculate the cost of carpeting. In this activity, the class will prepare a budget for redecorating the classroom. |

1.  First, the class decides what work needs to be done to redecorate the classroom. Some suggestions are listed below. Add to or modify this list, then assign each task to a group.

| Task | Group |
|---|---|
| Paint walls | |
| Paint ceiling | |
| Replace base trim | |
| Paint doors | |
| Replace door and window frames | |
| Install new carpet | |
| | |
| | |
| | |

When all tasks have been assigned to a group, one representative from the class should copy the table on the board.

●

2. As a group, think about your assigned task. Use the grid below to make a sketch of the appropriate geometric figure(s) you will be measuring.

Then, use a tape measure to measure all the required dimensions. Write down the measurements on your sketch, and be sure to include the units.

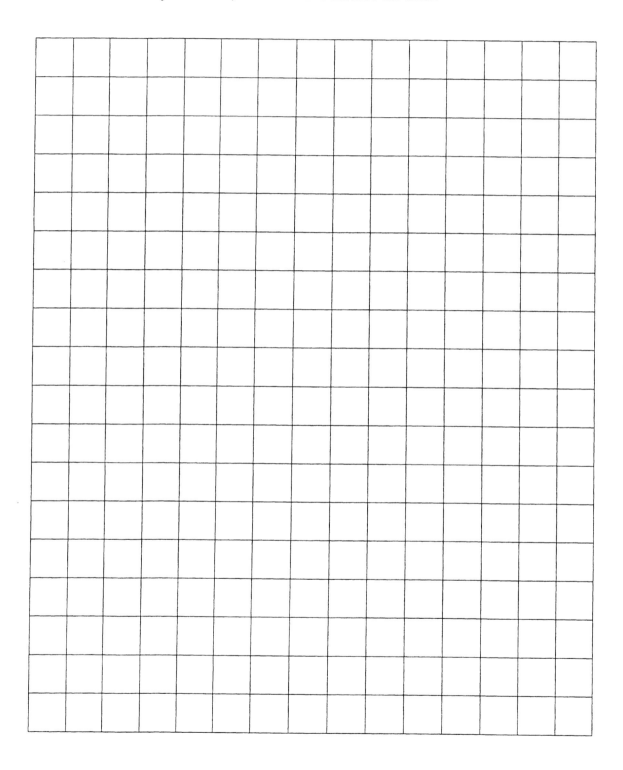

*Basic Mathematics*                    Collaborative Learning Activities

3.  Now, estimate the cost for your assigned task. Discuss whether you need to calculate the perimeter or area for your assigned task. If your group will be painting, visit or call a paint store to obtain the price of one gallon of paint. Do not include the cost for labor, as each group member will be donating time to this project! If your group will be installing new carpet, visit or call a carpet store to get the price for carpet installation.

    Write down your steps below. Show enough detail so that another group can follow your work.

    Exchange papers with another group. Analyze their work, and discuss with them any steps that are unclear.

4.  Write down the final cost estimate of your group's assigned task below, and also write the estimate next to the task on the board.

    Final Cost Estimate = $ _____

    When all groups have done this, add all the estimates to see how much it would cost to redecorate the classroom.

| Conclusion | This activity gave you a practical application of perimeter and area. As you can see, these concepts are used extensively in the field of redecorating, as well as other fields connected with the building trades. |
|---|---|

## Activity 9.2    Verify the formulas for the area of a parallelogram, triangle, and trapezoid.

| | |
|---|---|
| Focus | Areas of parallelograms, triangles, and trapezoids |
| Time | 10–15 minutes |
| Group size | 3 |
| Background | In Section 9.2 of your textbook, the formulas for the area of a parallelogram, triangle, and trapezoid are developed and used. This activity will show you a more elementary method for finding the areas, and provide a validation of the formulas. |

1.    Each group member will work with one of the three types of geometric figures. Write down the assignments below, then look up the formula for the area of your assigned figure. Write down the formulas in the appropriate space in the table.

| Figure | Group Member | Formula for Area |
|---|---|---|
| Parallelogram    A, B | | |
| Triangle    C, D | | |
| Trapezoid    E, F | | |

2.    As was explained in Section 9.2 of your textbook, the area of a region is equal to the number of square units that can be fitted into the region. Using this concept, count the number of squares inside the boundary of your assigned figures on the next page.

For squares that are not completely within the figure, estimate the fractional part of the square that is inside the boundary line. Combine fractional parts as accurately as possible to get the total number of squares. This will give you an approximation of the area of the figure. Write your results in the table on the following page.

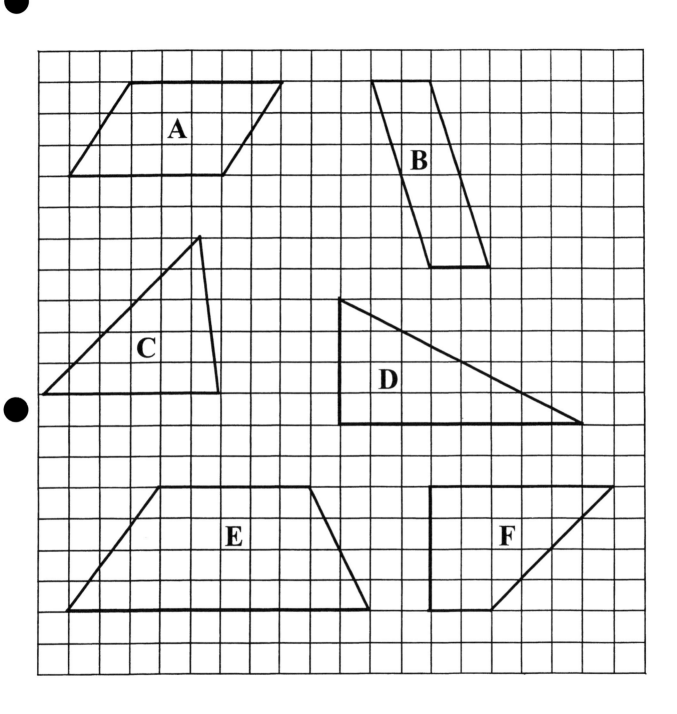

| Figure | | Area using squares | Base, b | Height, h | Area using formula |
|---|---|---|---|---|---|
| Parallelogram | A | | | | |
| | B | | | | |
| Triangle | C | | | | |
| | D | | | | |
| Trapezoid | E | | | | |
| | F | | | | |

3.  Now, measure the base and height of your assigned figures. Remember that each square of the grid counts as one unit. Write your results in the table above.

    Then, use the appropriate formula to calculate the area of each figure. Round your answers to the nearest whole number, and write your results in the table above.

4.  Compare the area you found using the formula with the area found by counting squares. How close are the values? Use complete sentences in your answer.

    Switch papers with a fellow group member, and check the work. How do the areas found using squares compare with the areas found using formulas?

    Exchange results with another group. Are their values comparable to yours?

| Conclusion | Using formulas gives an exact value for the areas of geometric figures. However, the method of counting squares can give you a quick estimate of the area. This method also gives you a deeper understanding of the meaning of area. |
|---|---|

# Activity 9.3    Estimate the value of π.

| Focus | Circles |
|---|---|
| Time | 10–15 minutes |
| Group size | 2–3 |
| Materials | String, ruler, calculator, soda can (optional) |
| Background | In Section 9.3 of your textbook, the circumference and diameter of a can are used to show that the constant, π, is equal to the ratio C/d. We will verify that this is true by calculating the values of C/d for circles of different sizes. |

1.  Use the string to measure the circumference of each circle on the next page. Place one end of the string anywhere on the circle, and carefully lay the string around the circle till you reach the starting point. Mark the ending point on the string, then use a ruler to measure the length of the string that is marked. Record this measurement in the table below.

    In addition, you may measure the circumference and diameter of any cylindrical object, like a soda can or roll of paper towels.

2.  Next, use a ruler to measure the diameter of each circle. Make sure you measure the diameter at the widest part of the circle. Record this measurement in the table also. Each group member should measure two or three items, while another group member writes down the measurements.

| Circle | Circumference, C | Diameter, d | $\dfrac{C}{d}$ |
|---|---|---|---|
| A | | | |
| B | | | |
| C | | | |
| D | | | |
| E | | | |
| F | | | |
| Soda can | | | |
| Towel roll | | | |

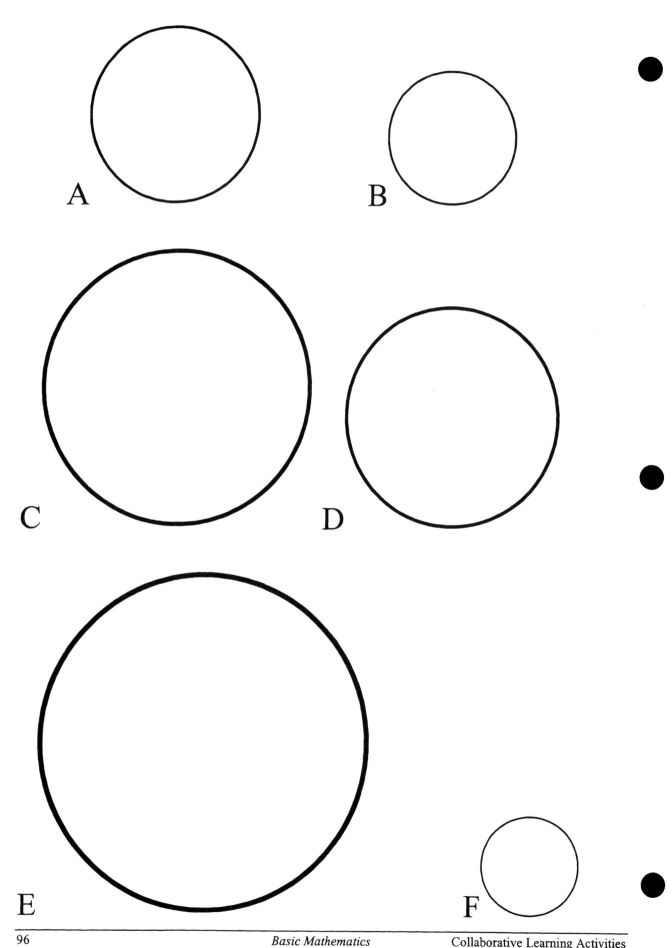

3. When you have completed all your measurements, use a calculator to compute the ratio C/d for each circle. Round your answers to 2 decimal places, and write the results in the table on the previous page.

4. Now, let's analyze the ratios C/d. Compare the values of C/d for all the circles. Are the ratios the same? Be sure to use complete sentences to answer all the questions in this step.

If not, how much do they differ by?

What could be the reasons for any differences?

Compare your values with those from another group. How close are your ratios?

What could be the reason for any discrepancies?

5. Calculate the average of the ratios, C/d, in your table, and write this value below.

$$\text{Average value of } \frac{C}{d} \ = \ \underline{\hspace{3in}}$$

How does your answer compare to that of the other groups in the class?

6. If you have time, find the average of the ratio $\frac{C}{d}$ for all the values calculated by each group in the class. How close is the class average to the value 3.14, or to a value for $\pi$ on a calculator?

| Conclusion | By measuring several circles, you have verified that the ratio of the circumference of a circle to its diameter is a constant. This ratio is called $\pi$, and is approximately equal to 3.14. |
|---|---|

## Activity 10.2    Add integers using a variety of methods.

| | |
|---|---|
| Focus | Addition of integers |
| Time | 20–25 minutes |
| Group size | 4 |
| Materials | Ruler |
| Background | A variety of methods is used to help students gain an understanding of addition of integers. Section 10.2 of your textbook shows addition on a number line and addition using rules. Another method that is quite effective is addition using tiles. This activity will give you practice with all three methods. |
| Instructor notes | Copy the page of tiles on card stock, and cut out the tiles. Each group will need one set of tiles. You can also purchase sets of color tiles from Cuisenaire Publications. If you choose to use color tiles, you will need to modify the instructions for step 3. |

1.    The three methods we will be working with are listed below.

Number Line Method

Tile Method

Rule Method

The following problems will be used to practice each of these three methods.

$$5 + (-7) = \underline{\hspace{1cm}}$$
$$-3 + (-9) = \underline{\hspace{1cm}}$$
$$-8 + 3 = \underline{\hspace{1cm}}$$

2.    Number Line Method

Draw a number line on a piece of ruled notebook paper. Turn the paper sideways and draw the line across the middle of the page. Mark the number 0 in the middle of the line, then mark the positive and negative integers. Use the ruled lines on the page to help space out the integers. Read the beginning of Section 10.2 of your textbook for the guidelines on using the number line. Remember that you move to the right if a number is positive, and to the left if the number is negative. If you like, show this on your number line by writing a "+" above the rightmost end of the line, and a "–" above the leftmost end.

To add 5 + (– 7), start at positive 5, and move _____ units to the left. You should end up at the point _____, so the answer is _____.

Practice with the other two examples. If your group needs more practice, create additional problems, or work some problems from your textbook.

3.    Tile Method

With the tile method, positive integers are represented by tiles with a "+" sign, while negative integer tiles have a "–" sign. When combining tiles, a positive tile and a negative tile add to 0, thus creating a "zero pair", which is removed.

Practice adding integers using this model. The first example is 5 + (– 7). Represent 5 with five positive tiles and – 7 with seven negative tiles. Put both groups together, and remove all zero pairs, and count the tiles remaining. In this case, there are _____ tiles, all _____, so the answer is _____.

Continue practicing addition with this tile model, until everyone in the group is comfortable with the process.

4.    Rule Method

Refer to Section 10.2 of your textbook for the different rules for adding integers. For reference, write a brief description of each rule in the table below. Be sure to use complete sentences.

| | |
|---|---|
| Positive + Positive | |
| Negative + Negative | |
| Positive + Negative  OR  Negative + Positive | |

Now, do the example problems using these rules. As before, practice adding integers until each group member is proficient with the method.

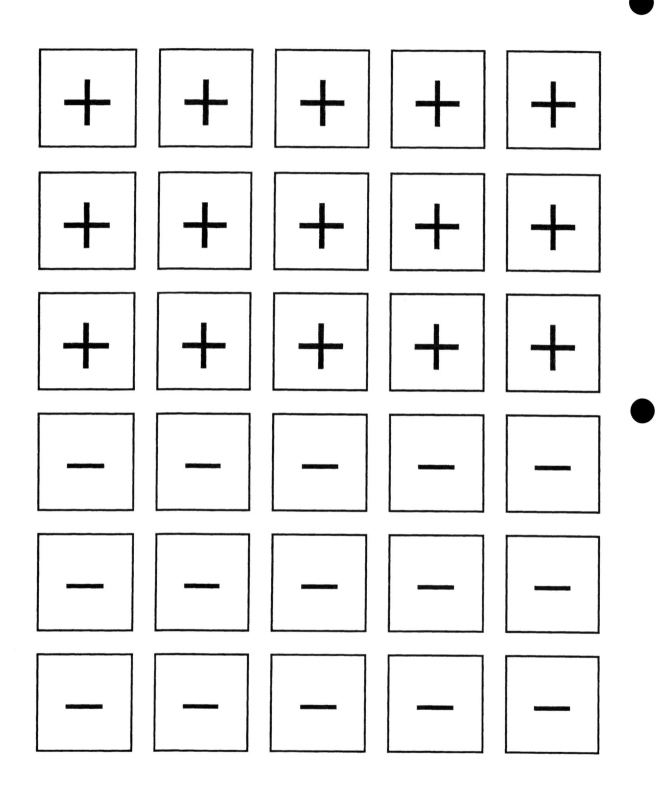

5.  For the second part of this activity, each group member chooses one of the addition methods. The fourth group member takes the role of the question generator.

    To begin, the question generator creates an integer addition problem. The problem can have up to five numbers, and should contain a mix of positive and negative integers. Keep the values between – 15 and + 15, to accommodate the model restraints.

    Write the problem down on a piece of paper so the other group members can clearly see the numbers. Then read out the problem slowly while the other group members work the problem using their chosen method. When all group members are finished, compare answers. If the answers do not match, check each other's work, and redo the problem as needed.

    Continue in this way two more times, with the question generator creating 2 more problems. Then, rotate roles so that each group member takes a turn as the question generator, and works with each of the methods.

| Conclusion | This activity shows you three ways of representing the addition of integers. Each method has advantages and disadvantages. While the number line and tile methods are very useful for showing the logic behind integer addition, they are not practical for adding integers that are large. The rule method, on the other hand, works well for integers of any size. However, memorizing rules without understanding the reasons behind them is dangerous, and can lead to difficulties later on in your study of algebra.<br><br>Ideally, you would gain the understanding of integer addition through the use of the number line or tiles, but use the rules for doing the actual addition. |
|---|---|

## Activity 10.3   Subtract integers using tiles.

| | |
|---|---|
| Focus | Subtraction of integers |
| Time | 15–20 minutes |
| Group size | 2 |
| Materials | Scissors |
| Background | We subtract integers by adding the opposite of the number being subtracted, as described in Section 10.3 of your textbook. On a more basic level, subtraction corresponds to "take away," as was described in Section 1.3 of your textbook. In this activity, "take away" will be used to subtract integers when working with tiles. |
| Instructor notes | Copy the next page on card stock, and cut out the tiles. Each group will need two pages worth of tiles. You can also purchase sets of color tiles. If you choose to use color tiles, you will need to modify the instructions for step 1. |

1.   Note: If your group has done Activity 10.2, and is proficient with using positive and negative tiles to add integers, skip ahead to step 2.

    With the tile model, positive integers are represented by tiles with a + sign, while negative integer tiles have a –sign. When combining tiles, a positive tile and a negative tile add to 0, thus creating a "zero pair," which is removed.

2.   For this activity, one group member holds all the positive tiles, while the other keeps the negative tiles. Practice combining positive and negative tiles. Each group member places several of his or her tiles on the table. It does not matter how many of each you select. Together, remove all zero pairs, and count the remaining tiles left on the table. There should be only one kind of tile. Practice combining tiles until both of you are proficient with this process.

3.   Now, let's see how subtraction is done using the tiles.

    Consider the problem            $9 - 5 =$ _____.

    Represent 9 with nine positive tiles. Since subtraction means "take away," you will remove five positive tiles from the group. The tiles that are left represent the answer to the problem. Next, look at the problem    $-7 - (-2) =$ _____.

    Start with seven negative tiles and take away two negative tiles. The answer is represented by the remaining tiles.

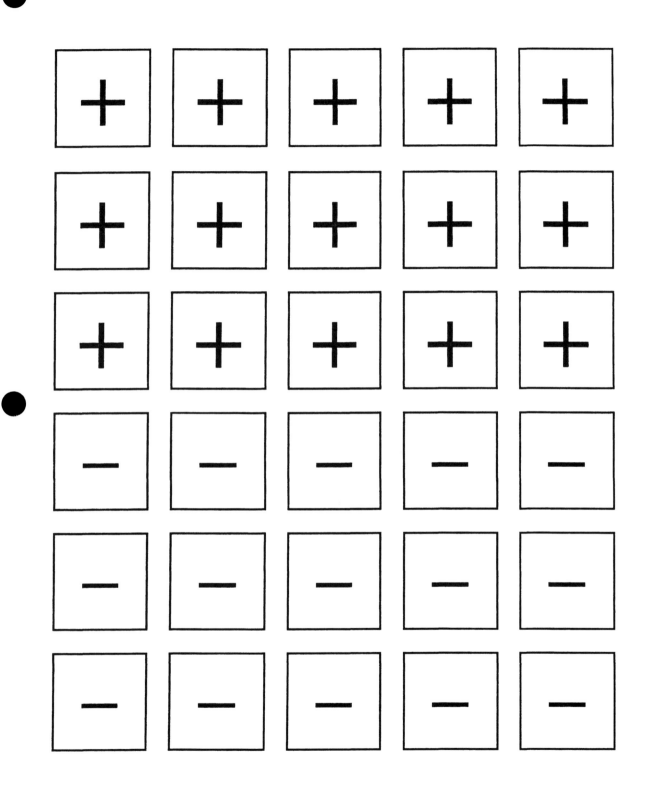

Practice this process on the following problems. Take turns, so each group member gets to work with the model.

$$7 - 4 \ = \ \underline{\hspace{2cm}}$$

$$-9 - (-5) \ = \ \underline{\hspace{2cm}}$$

$$-6 - (-4) \ = \ \underline{\hspace{2cm}}$$

4.  Once you are both proficient with the problems in the previous step, you can move on to more complex situations. Consider the problem

$$3 - 8 \ = \ \underline{\hspace{2cm}}.$$

As before, start with three positive tiles, and try to take away eight positive tiles. This is not possible, since there are not enough positive tiles to remove. To resolve this difficulty, add more positive tiles until there are enough to take away. The other group member will need to add an equal number of negative tiles so that the problem is not altered. Remember that adding zero pairs is equivalent to adding the number 0. Remove eight positive tiles, and count the remaining tiles to get your answer.

Now, work together and use the color tile method to do the following subtraction problems.

$$2 - 9 \ = \ \underline{\hspace{2cm}}$$

$$1 - 7 \ = \ \underline{\hspace{2cm}}$$

$$-3 - (-8) \ = \ \underline{\hspace{2cm}}$$

$$2 - (-4) \ = \ \underline{\hspace{2cm}}$$

$$-6 - (-6) \ = \ \underline{\hspace{2cm}}$$

$$-1 - 7 \ = \ \underline{\hspace{2cm}}$$

$$3 - 5 \ = \ \underline{\hspace{2cm}}$$

$$3 - (-5) \ = \ \underline{\hspace{2cm}}$$

$$-3 - (-5) \ = \ \underline{\hspace{2cm}}$$

$$-3 - 5 \ = \ \underline{\hspace{2cm}}$$

| Conclusion | The color tile method gives you a visual, hands–on model for integer subtraction. Use this model to help you understand why the rules for subtraction work, and to develop greater proficiency in subtracting integers. |
| --- | --- |

## Activity 10.5   Use the order of operations as a group to simplify expressions.

| Focus | Order of operations |
|---|---|
| Time | 20–30 minutes |
| Group size | 3 |
| Background | Simplifying expressions using the rules for order of operations can be quite confusing for complicated expressions. Learning to simplify expressions as a group will help clarify the process. |

Rules for Order of Operations

| | | Do all calculations within parentheses before operations outside. |
|---|---|---|
| | **E** | Evaluate all exponential expressions. |
| | **MD** | Do all multiplications and divisions in order from left to right. |
| | **AS** | Do all additions and subtractions in order from left to right. |

1.  Before you begin simplifying expressions, study the rules for order of operations above. Assign each group member to one of the steps listed. Write the name of the group member next to his or her assigned task in the table above. Note that the first step (calculations within parentheses) is not assigned. All group members will do this step together.

2.  NOTE:  If your group has done Activity 1.9, 3.7, or 4.4, and is familiar with the group method for simplifying expressions, you may skip ahead and simplify the expression on the next page.

    Now you are ready to simplify expressions as a group. Analyze the expression together and decide on the first step. If there are parentheses, decide whether the expression inside the parentheses needs to be simplified. Following the order of operations, **E** will perform his or her task before **MD**, and **MD** will perform his or her task before **AS**.

    Practice on the example on the next page. (This is Example 21, Section 10.5 in your textbook.) The first step has been done for you: Multiply inside the parentheses. **MD** will do this step, writing "**MD**" in the left box, and writing the new expression below the original expression. Continue simplifying the expression by passing the problem to the appropriate group member for the next step. When you are done, compare your steps to those in Example 21, Section 10.5 in your textbook. If there are any discrepancies, discuss them within your group. Compare your result with the other groups. Are they the same? Discuss any differences with the other groups.

Example 21, Section 10.5

| | |
|---|---|
| | $2^4 + 51 \cdot 4 - (37 + 23 \cdot 2)$ |
| **MD** | $2^4 + 51 \cdot 4 - (37 + 46)$ |
| | |
| | |
| | |
| | |
| | |
| | |

3.  Once you understand the process, choose an expression from problems 51 – 88 in Exercise Set 10.5 in your textbook to simplify as a group. Use the table on the next page to organize your work. Make as many copies as you need. Alternatively, you can draw the table on a blank sheet of paper. Do as many problems as you can in the time allotted.

| Conclusion | This activity should help you gain a better understanding of the rules for order of operations. You can also use this method in the future, when you simplify algebraic expressions. |
|---|---|

● Original expression _____

## Activity 11.4    Create and solve equations as a group.

| Focus | Solving equations |
|---|---|
| Time | 20–25 minutes |
| Group size | 4 |
| Background | Solving equations is a very important skill in algebra.  This activity will give you practice in using the addition and multiplication principles to solve equations. |

1.    Each group will create and solve equations by following the steps outlined below.  Study the example so you understand the mechanics of this process.

|  |  | Example |
|---|---|---|
| Step 1 | The first group member writes $x$ = some number on a piece of notebook paper, and passes the paper to the second group member. | $x = 4$ |
| Step 2 | The second group member multiplies both sides of the equation by a number.  Write the new equation below the first one, and pass the paper to the third group member. | $3x = 12$ |
| Step 3 | The third group member adds or subtracts a number to both sides of the equation, and writes the new equation below the second one.  Fold over the paper so that only the last equation is shown, then pass the paper to the fourth group member. | $3x - 5 = 7$ |
| Step 4 | The fourth group member writes the equation on his or her record sheet, and solves the equation.  When done, unfold the notebook paper and check that the solution matches the equation written by the first group member.  If it does not, the group should find out where the mistake occurred. | $3x - 5 = 7$ <br> $3x - 5 + 5 = 7 + 5$ <br> $3x = 12$ <br> $\dfrac{3x}{3} = \dfrac{12}{3}$ <br> $x = 4$ |

# RECORD SHEET

**ROUND 1**      Equation:

Solution:

**ROUND 2**      Equation:

Solution:

**ROUND 3**      Equation:

Solution:

**ROUND 4**      Equation:

Solution:

2. Now you are ready to create your own equations. In Round 1, use only positive integers in steps 1, 2, and 3.

   Each group member begins with step 1, writing down the equation on a piece of notebook paper. Pass the paper to the group member on your right.

   Modify the equation you receive according to step 2, then pass the paper to your right again. Continue until you reach step 4; you should have an equation that looks like $ax + b = c$. Write down this equation on the record sheet on the previous page. Then, solve and check the equation, following the directions for step 4.

   When all group members are done, examine all four equations in your group and correct any errors.

3. For Round 2, you may use negative integers in steps 1 and 2. Start with a fresh sheet of notebook paper and pass the equation around the group as before.

4. Round 3 gets a little more complicated, as you may use a fraction when multiplying in step 2.

5. Finally, in round 4, you may use a fraction in steps 1, 2, or 3.

| Conclusion | Notice that you are writing equivalent equations in steps 2 and 3. This is one of the fundamental concepts in equation solving, as explained in Section 11.2 of your textbook. This activity should give you a clearer understanding of how the addition and multiplication principles work, and consequently increase your proficiency in solving equations. |
|---|---|